SpringerBriefs in Applied Sciences and Technology

PoliMI SpringerBriefs

Springer, in cooperation with Politecnico di Milano, publishes the PoliMI Springer-Briefs, concise summaries of cutting-edge research and practical applications across a wide spectrum of fields. Featuring compact volumes of 50 to 125 (150 as a maximum) pages, the series covers a range of contents from professional to academic in the following research areas carried out at Politecnico:

- Aerospace Engineering
- Bioengineering
- Electrical Engineering
- Energy and Nuclear Science and Technology
- Environmental and Infrastructure Engineering
- Industrial Chemistry and Chemical Engineering
- Information Technology
- Management, Economics and Industrial Engineering
- Materials Engineering
- Mathematical Models and Methods in Engineering
- Mechanical Engineering
- Structural Seismic and Geotechnical Engineering
- Built Environment and Construction Engineering
- Physics
- Design and Technologies
- Urban Planning, Design, and Policy

http://www.polimi.it

Marika Fior · Paolo Galuzzi · Gabriele Pasqui ·
Piergiorgio Vitillo

(Re)Discovering Proximity

Generating New Urbanity—An Action Research for Milan

POLITECNICO
MILANO 1863

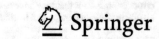

Marika Fior 🆔
Department of Planning, Design
and Technology of Architecture
Sapienza University of Rome
Rome, Italy

Paolo Galuzzi 🆔
Department of Planning, Design
and Technology of Architecture
Sapienza University of Rome
Rome, Italy

Gabriele Pasqui 🆔
Department of Architecture and Urban
Studies
Politecnico di Milano
Milan, Italy

Piergiorgio Vitillo 🆔
Department of Architecture, Building
Engineering and Building Environment
Politecnico di Milano
Milan, Italy

ISSN 2191-530X   ISSN 2191-5318 (electronic)
SpringerBriefs in Applied Sciences and Technology
ISSN 2282-2577   ISSN 2282-2585 (electronic)
PoliMI SpringerBriefs
ISBN 978-3-031-08957-2   ISBN 978-3-031-08958-9 (eBook)
https://doi.org/10.1007/978-3-031-08958-9

This Springer imprint is published by the registered company Springer Nature Switzerland AG
The registered company address is: Gewerbestrasse 11, 6330 Cham, Switzerland

Preface

The book is the result of reflections and research in the field of urban planning and urban design that have dealt with the issue of proximity in recent years. After the pandemic, many scholars have tried to reinterpret the urban dimension. Many have noted the fragility of cities (technologically, environmentally, and socially). In contrast, in this book, we have chosen to highlight the potential of cities to cope with economic uncertainties and the global challenges affecting the planet. The aim is to bring urban design back to the centre of the debate, especially the public space design. The objective is to translate policies and strategies into concrete actions to improve the habitability of living environments by re-establishing a connection between space and society.

To write this book, we drew on research conducted at the Politecnico di Milano between 2018 and 2019. The research aimed to rethink the role of urban space after the introduction of the new M4 metro line in Milan. The research findings were the driving force to show how the reshape of urban space can unlock the adaptive and transformative capacity of cities by meeting both the socioeconomic demand to increase welfare and services for the population and the environmental demand for improving the ecological and health performance of historic and densely built settlements.

The book consists of five chapters.

Chapter 1, an introduction, describes the current relationship between people and spaces, society and places, and proximity and distance. There is a general reference to the need to reduce social inequalities and improve access to the city.

Chapter 2 looks at proximity as a way of reconfiguring the contemporary European city (Milan in particular) based on its historical structuring into neighbourhoods. The study of the Milan underground has allowed us to highlight the relationship between the M4 and the identity of the places it crosses, the different neighbourhoods, by proposing the metro stations as real neighbourhood hubs.

Chapter 3 constructs a palimpsest for designing the interferences between the M4 metro stations and the historical, social, and landscape values that play a role in the redesign of the stations in a perspective of resignification of the city and its heritage.

Chapter 4 assumes the vocation of the new infrastructure as a field for experimenting with new practices in sustainable mobility based on walkability and a new green and blue connections. This perspective allows us to imagine the stations and the M4 as a contemporary urban ecological corridor in the European view of environmental transition.

The final Chap. 5 describes the technical instruments utilised (the master plan and the multi-level spatial strategies and operative guidelines for urban design at the scale of proximity) used to redesign the spaces of the M4 line on the surface of the city as a platform for enabling urban regeneration.

Therefore, this book intends to offer suggestions and operative tools whose use could be extended from Milan to other cities and infrastructural operations planned for different urban areas of Italy and Europe.

Authorship

The authors agreed on the structure and contents of the book. In particular, Marika Fior wrote Chaps. 3 and 4, Paolo Galuzzi wrote Chap. 5, Gabriele Pasqui wrote Chap. 1, and Piergiorgio Vitillo wrote Chap. 2.

Acknowledgements

The book is part of the 2018–2019 DAStU research and its outcome.

Paolo Beria, Lucìa Bocchimuzzi, Marika Fior, Paolo Galuzzi, Francesco Infussi, Antonio Longo, Laura Montedoro, Filippo Oppimitti, Gabriele Pasqui, Laura Pogliani, Paola Pucci, and Piergiorgio Vitillo developed the research.

The City of Milan, MM Spa Direzione Comunicazione (Milan), and AMAT Srl (Milan) collaborated and supported the research.

Rome, Italy	Marika Fior
Rome, Italy	Paolo Galuzzi
Milan, Italy	Gabriele Pasqui
Milan, Italy	Piergiorgio Vitillo

Contents

List of Figures

Chapter 1
Proximity, Distance, Urban Space and the Human Body: An Introduction

Abstract The chapter contains a general introduction to the topics of the book against the backdrop of a reflection on the notions of proximity, distance in urban space and the effects of the pandemic. On the one hand, the text focuses on the bodily dimension of how we experience the city. On the other hand, it looks at the link between the construction of proximity spaces and the attempt to contrast social and spatial inequalities. The general objectives of the book and its structure are defined in this context.

Cities have long been places for experimenting with forms of proximity and practices of distancing. On the one hand, urban settlements have been characterised by the density of human bodies and relations from their most distant origin. They are places where diverse activities and functions and different social groups come together in a shared space. On the other hand, owing precisely to this plurality of sites, functions and people, cities present multiple forms of distancing, internal boundaries, enclosures and thresholds, which divide activities and populations.

Proximity and distance are thus two constituent dimensions that coexist in urban space: we cannot be close without assigning a scale to distance, intended primarily as the threshold between bodies and the possibility for fertile encounters between different people.

The Covid-19 pandemic, if we think carefully, induced us to think above all in terms of the dialectic between constrained proximity (but not cancelled, as this would be impossible) and distance, challenging both architecture and urbanism to occupy a threshold between these two poles.

This twofold nature of urban space, this continuous oscillation between density and distance, proximity and separation, has become the fundamental trait of the contemporary city. As Ash Amin tells us, it is increasingly more a "land of strangers" (Amin 2012), where diverse populations with little in common share spaces (beginning with public ones) in a continuous renegotiation of possibilities to encounter one another. As Louis Wirth was well aware, the simplest definition of the city refers precisely to this density, potentially conflictual, between different people: "For sociological purposes a city may be defined as a relatively large, dense and permanent settlement of socially heterogeneous individuals. Based on the postulates which this minimal definition suggests, a theory of urbanism may be formulated in the light of existing knowledge concerning social groups" (Wirth 1938: 8).

This heterogeneity has brought new forms of proximity concerning the growing pervasiveness of information technologies and communication to the contemporary city. Never as now, and never as during the Covid-19 pandemic, have we experienced the relevance of what Melvin Webber referred to as "communities without propinquity" (Webber 1963). However, this in no way signifies, as we shall amply demonstrate in this book, a loss in the importance of physical proximity: instead, it weaves together diverse forms of relations, some remote, mediated for example, by social media and more comprehensively by the network in which we are all increasingly enmeshed.

In other words, proximity, both material and virtual, alludes to the contemporary city and the complex issue of a possible coexistence, in a context witness to the radical pluralisation of forms of life (Pasqui 2018). Furthermore, proximity alludes to the difficulty in recognising principles of coexistence based on references to pre-constituted identities and communities.

The simultaneous presence of proximity and distance is thus a principal ingredient of the city's essence. What is more, it also depends on the fact that, as human beings, we experience the city through the physicality of our bodies. There is more: bodies involved in practices of proximity are also 'theatrical bodies' that act in urban space as if on stage. Through encounters, through the forms of social interaction that characterise public life, these 'theatrical bodies' follow different scripts, some mandatory, some free and spontaneous, forming the meaning of urban spaces and transforming spaces into places.

The relations that occur in public, to cite Erving Goffman, are an integral part of life in the city, of that "everyday life as representation" that defines how we inhabit places and social and spatial relations, the "rituals of interaction" that connote and constitute them (Goffman 1959).

We are thus human bodies that encounter the body of the city. We have always known this, and we have verified it with particular urgency during the now lengthy time during which the pandemic forced us to remain separated and isolated. Reflecting on the link between the pandemic and the city means reflecting on spaces and bodies, along with a debate on relations between space and society that is anything but simplistic (Pasqui 2022).

If we adopt a Spinozian perspective, each body is a pattern, an extremely complex set of bodies composed in accordance with multiple relations. The human body is organised according to variable and shifting links traversed by a myriad of other bodies. The bacteria that inhabit our body, that compose it, are as heavy as our brains. The virus that changed our lives is merely another body that decomposes essential relations within our bodies. In short, an unpleasant encounter.

Similarly, the city is an assemblage of bodies: buildings, open spaces, streets and squares, infrastructures, technological services, plants, animals, clean air and suspended particles, networks and information. Bodies create variable compositions between one another and with our bodies. Bodies that are more or less malleable and more or less porous: once again, good encounters and unpleasant encounters.

This is how we inhabit urban space: encountering other bodies, some presenting resistance, others offering purchase, composing our body within the space it occupies, in its bottleneck, its plurality, with urban bodies and the multiple populations that cross and use the city (Pasqui 2008). When an encounter is good, our strength is increased; when it is unpleasant, it is diminished.

How then are we to imagine the design of urban space, particularly proximity spaces and sharing, within a similar perspective? How are we to imagine the action of a project when we intend primarily to foster positive encounters? What can we do to simplify positive encounters, knowing that not everything depends on us and that unpleasant encounters, tripping hazards, violence, viruses and catastrophes remain possible and largely unpredictable?

A reflection on the proximity spaces begins with the need to discover our existence in urban space, in public, in front of others. We must describe and imagine our ordinary and everyday habits of encountering others and coexisting in the spaces of the city, knowing it is above all in public proximity spaces that the quality of our encounters can improve, that potential conflicts can find composure here.

How then are we to imagine spaces open to what is possible; spaces that do not bridle and regulate the body but offer footholds and occasions for practices of proximity? The first step is precisely the construction of a different way of thinking about how we design public space. It is a question of imagining public space as a space that corresponds with the moment and with the injunction of the plurality of forms of life, with flexible space. If we seek a correspondence with this moment, we must imagine public space as a space open to many different uses, to unpredictable possibilities. The concept of 'openness,' to unforeseen events, nonetheless presents different shades. An 'open' space could initially be intended as versatile: an object is versatile when it can be used in many different ways, when it lends itself, in virtue of its conformation, performance and structure, to diverse uses. A 'versatile' space permits only certain possibilities for use, but obviously excludes others. On the other hand, open space can be 'vague', undefined, in the sense it does not strictly prefigure any particular use, as it lacks complete and designed qualities. Vagueness and versatility have the potential to be contradictory. To be versatile, a space may need to be heavily designed and planned; while a space can be vague even when it is weakly designed, or not planned at all. What equilibrium must architects identify when playing with the line between versatility and vagueness, leaving room for that openness that favours the freedom and innovativeness of practices? How are we to interpret this moment in design, knowing that in any case the body requires footholds and resistances in order for it to inhabit urban space and to dance within it? How does the gap between vagueness and versatility, between suspension and active modification through design, permit us to offer the body, literally, that openness that allows for a possible coexistence? Perhaps coexistence needs to be imagined in the sense of the *cum*, what the Italian philosopher Carlo Sini called "co-possible aggregation" (Sini and Pasqui 2020).

There is another aspect to consider: the question of proximity is not politically neutral because it has solid ties with social justice. The reflection on the 15-min city—developed during the pandemic though rooted in thinking on the disciplines of urban planning—demonstrates the need to reinvent proximity spaces in a context capable of radically reducing socio-spatial inequalities, the gaps between parts of the city, where services and facilities and the performance of open public spaces are severely asymmetrical.

An example: the most drastic effects of the pandemic have been suffered by those living in poor conditions and those with access to limited or poor quality open spaces. In substance, the pandemic only radicalised the forms of socio-spatial inequality affecting all cities in Italy and Europe, within the more general framework of the inequalities that have characterised the Western world over the past fifty years (Piketty 2015).

In this context, design has become decisive to the construction of proximity spaces which assume this complexity and this plurality. Spaces capable of supporting possibilities for fertile interaction among diverse opportunities for socialisation and platforms for contrasting spatial injustices.

These conceptual frameworks support the strength that returns in this book, which proposes a rediscovery of the design of proximity spaces, beginning with what we have learned and continue to learn from the terrible pandemic experience, and from a concrete example of urban design.

Once again, this is not a new topic: the tradition of urban planning offers multiple examples for the study of proximity spaces. Nonetheless, it seems evident that the pandemic urges us to change our point of view and undertake new experiments in design.

The effects of the Covid-19 pandemic, and several consequent reflections on the future of cities (Cannata 2020; Nuvolati and Spanu 2020; Ardenne and Femia 2021), have recentred interests on the concept of proximity (Manzini 2021; Pellegrini 2012). Proximity intended as both the relationship between communities and urban functions, and as relations among people, built spaces and open spaces. An historic and fertile field of interest for Anglo-Saxon and Northern European urban studies (Giedion 1941; Cullen 1961), manifestly represented in Italy by the policies and resulting neighbourhoods of the INA Casa[1] (Di Biagi 2001; Pilat 2019), a spatial and social programme that appeared to have been surpassed by the styles and rhythms of life in the contemporary city. In parallel, an 'action research' developed by the Department of Architecture and Urban Studies (DAStU) at Politecnico di Milano[2] reached its conclusion. The research concentrated on contextualizing the stations of the new M4 metro line to produce a proposal that would join flows with places and long-range networks with local ones. The goal hinged on the idea of transforming

---

[1] The INA Casa programme was a post-war housing plan promoted by the Italian State to build some two million units. It takes its name from the *Istituto Nazionale delle Assicurazioni*, National Insurance Institute, which funded the programme.

[2] The research, entrusted to the DAStU by the City of Milan and Metropolitane Milanesi SpA, was articulated in two phases (2017–2018 and 2019) and developed by a Research Group coordinated by Gabriele Pasqui and composed of Paolo Beria, Paolo Galuzzi, Francesco Infussi, Antonio A. Longo, Laura Montedoro, Laura Pogliani, Paola Pucci, Piergiorgio Vitillo; and by an Operative Unit composed of Lucia Bocchimuzzi, Marika Fior, Filippo Oppimitti.

these new stations into regenerative urban thresholds, platforms for enabling environmental systems, systems of settlement and infrastructures in the neighbourhoods of the city crossed by the new metro line (Fior et al. 2019).

When the pandemic focused attention on the core theme of this reconsideration of the stations of the new M4 line and their context, at different scales, it assumed new meanings concerning the unprecedented situation raised by Covid-19 for anyone dealing with urban spaces and services. Returning to the consideration of the new metro stations in the wake of the pandemic makes it possible to relaunch a reflection anchored to the design of proximity spaces.

This book has been written to foster a return to considering the theme of proximity and the relations it generates in the urban fabrics of the contemporary city by borrowing the contents and results of the aforementioned research. The theme of neighbourhoods returns powerfully in interpretations of the contemporary city after last appearing during the 1950s and '60s as an alternative to sprawling development and as a representative, though not exclusive, component of public initiative. That said, even the city's quarters face social, demographic and spatial decomposition and processes of fragmentation, which any urban project must be able to incorporate and deal with.

On the other hand, large infrastructures, such as Milan's new M4 metro line, (Fig. 1.1) are also an occasion for rethinking the city in its entirety. The possible connections between open spaces and systems of landscaping lining new Green and Blue Infrastructures (GBI), assuming the plural identities of the urban space traversed by this infrastructure, as an important platform for redesigning the proximity spaces.

Substantially, the study related to the M4 offered a privileged vantage point for a broader and more articulated reflection on proximity-centred design strategies for urbanism.

The thesis supported here is that the challenge of restarting immediately after the pandemic by focusing on a new and different urbanity is difficult though possible. However, only under the condition of coherence and integration of strategies and actions in the programming of urban transformations, coupled with innovation in the field of regulations, and an interweaving of issues of welfare, public space, active communities, slow connections, health and urban safety, the environment, and sustainable tourism.

The overarching objective of this book is the proposition of a vision of contemporary dwelling based on a renewed form of proximity and the values it brings. Values that, on the one hand, consider the need to build more just cities, in which services and facilities are distributed across the city and, on the other hand, the need to reimagine proximity spaces considering the complex dialectic between distance and new practices of sharing.

References

Amin A (2012) Land of strangers. Polity Press, London

Ardenne P, Femia A (2021) La buona città per una architettura responsabile, Marsilio, Venice

Cannata M (ed) (2020) La città per l'uomo ai tempi del Covid-19. La Nave di Teseo, Milan

Cullen G (1961) Townscape. Architectural Press, London

Di Biagi P (ed) (2001) La grande ricostruzione. Il piano Ina-Casa e l'Italia degli anni cinquanta, Donzelli Editore, Rome

Fior M, Galuzzi P, Vitillo P (2019) Metro M4, the new green-blue backbone of Milan. From infrastructure design to urban regeneration project. In: XIII CTV 2019 proceedings: XIII international conference on virtual city and territory: 'challenges and paradigms of the contemporary city', p 8433–8448, Barcelona

Giedion S (1941) Space, time and architecture. The growth of a new tradition, Cambridge. It. tran. 1954, Spazio tempo e architettura, lo sviluppo di una nuova tradizione, Hoepli, Milan

Goffman E (1959) The presentation of self in everyday life. Anchor Books, New York

Manzini E (2021) Abitare la prossimità. Idee per la città dei 15 minuti, Egea, Milan

Nuvolati G, Spanu S (eds) (2020) Manifesto dei sociologi e delle sociologhe dell'ambiente e del territorio sulle città e le aree naturali del dopo Covid-19. Ledizioni, Milan

Pasqui G (2008) Città, popolazioni, politiche. Jaca Book, Milan

Pasqui G (2018) La città, le pratiche, i saperi. Donzelli, Rome

Pasqui G (2022) Coping with pandemic in fragile cities. Springer, Berlin-Milan

Pellegrini P (2012) Prossimità. Declinazione di una questione, Mimesi Edizioni, Sesto San Giovanni (MI)

Piketty T (2015) The economics of inequality. Harvard University Press, Cambridge (US)

Pilat SZ (2019) Ricostruire l'Italia. I quartieri Ina-Casa del dopoguerra, Castelvecchi Editore, Rome

Sini C, Pasqui G (2020) Perché gli alberi non rispondono. Lo spazio urbano e i destini dell'abitare. Jaca Book, Milan

Webber M (1963) Order in diversity: community without propinquity. In: Wingo LJ (ed) Cities and space. Joan Hopkins Press, Baltimore, pp 23–56

Wirth L (1938) Urbanism as a way of life. Am J Sociol XLIV(1):1–24

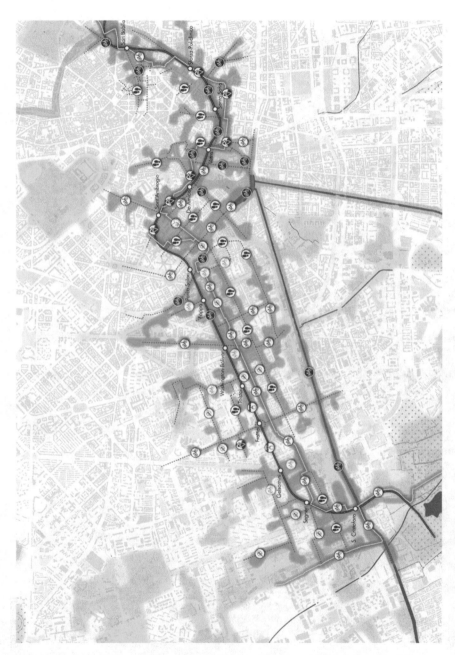

Fig. 1.1 The green–blue backbone of the M4: master plan of project actions. Original drawing at 1:150,000

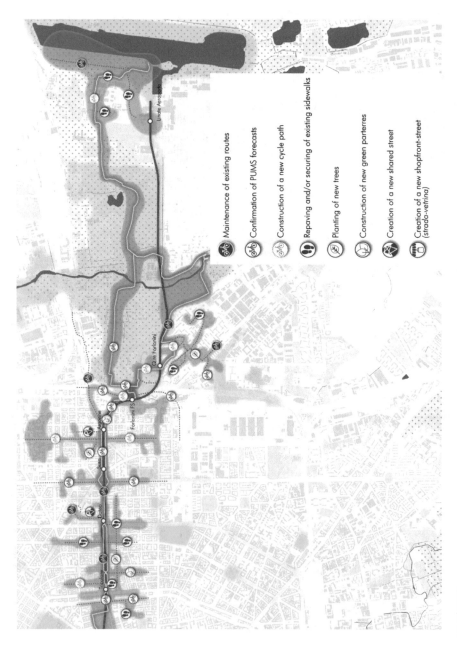

Fig. 1.1 (continued)

Chapter 2
Designing Neighbourhoods to be Called by Name

Abstract In this chapter, proximity is defined as an opportunity for redesigning the contemporary European city (in particular Milan) based on its intelligent historical structuring in neighbourhoods. This concept has deep roots in the history of the city, and links to intense urban transformations and the urban metabolism of the city over time. It also regards urban transformations and regeneration projects at diverse levels of granularity (city, quarter and building). Finally, it is intertwined with pedestrian movement and public mobility being imposed by new lifestyles and the necessary ecological transition to be promoted together with local communities.

The chapter's four paragraphs look at the following aspects. The first sheds light on how the Covid-19 pandemic offers the city a chance to rethink its urban spaces, both open and built, as a system of 'archipelagos of neighbourhoods,' each with its own identity learning from the qualitative characteristics and values inherited from the historical city. The second paragraph begins with the transformations of the existing city, an organism that changes slowly but incessantly, driven by an extensive and spontaneous metabolism. It reveals how the urban structures of the nineteenth century, notably Cesare Beruto's plan for Milan (1889), have proven resilient and capable of adapting and improving, manifesting ductility and resistance to processes of reuse, concerning typologies and forms of settlement and the offering of services. The third paragraph focuses on the regeneration of existing building stock as an indispensable policy for improving urban inhabitability. In particular, this means updating the dimensions and characteristics of housing to reflect new social morphologies and contemporary lifestyles; learning to work incrementally with the existing by restitching and mending; and expanding the field of interventions to relational aspects and the bond with the existing and consolidated city. The fourth paragraph returns to selected features of the project to contextualise the stations of Milan's M4 metro line (the Green and Blue Backbone project). Specifically, the paragraph highlights the possibility to transform the nodes of the new metro stations into community 'hubs,' enriched by integrated and connected public functions that contribute to the creation of a city at the scale of active mobility.

2.1 Inhabiting Proximity

2.1.1 Form Global (Pandemic) to Local (Life)

European cities are a palimpsest of overlaps, juxtapositions, and stratifications in perpetual movement and modification, whose dynamism is tempered by their geographic and historic nature. Italy was initially a land of small settlements, even after processes of industrialisation sanctioned the passage from the ancient to the modern city, and to today's contemporary city.

The Covid-19 pandemic, which has revealed our fragilities, may be an occasion for rethinking urban space, both open and built. It can challenge our capacities to build the world, not as an end ("Never let a good crisis go to waste" as Churchill stated at the end of the Second World War), but by maximising urban capacity. The 'capability approach' (Blečić et al. 2013) is the ability to welcome a multitude of forms and practices, including and mixing the great variety that makes it so rich. Numerous different voices participate in constructing a polyphonic portrait, an articulated and open weave. This process involves neighbourhood communities and networks, once forgotten and later rediscovered during the difficult phase of the lockdown.

Despite several pessimistic reflections on the fate of the metropolis (Benevolo 2011), the Covid-19 pandemic will not wipe out cities, though it did broadly expose their fragilities and vulnerabilities. Health emergencies have defined the history of the world's cities, which have almost always managed to offer appropriate responses to the pandemics that beset them (from the Plague of Athens during the fifth century BC to the Black Plague that devastated Europe during the fourteenth century, to the epidemics of Victorian London). These events failed to cancel the model of urban life, and also became occasions for and accelerators of responses and technological and socio-environmental solutions. One of the reasons why cities have proven resilient is that their spaces, infrastructures and social practices have demonstrated an ability to adapt rapidly.

Cities make it possible to create unique social, cultural, and economic networks of relations among people, which are inherent characteristics of the urban world. These relations are based on the social interaction favoured by proximity, contact between neighbours, face-to-face experience (a true and proper technology of communication), and the discovery of the unexpected (serendipity). All of which permits an exchange of knowledge that cannot be codified or transmitted remotely. Direct encounters and personal exchanges stimulate active participation and imply mutual commitment; there is no collective growth without an authentic sense of social interaction, yet another reason that cities are not destined to die.

This awareness suggested constructing the city as an 'archipelago of neighbourhoods,' an experience that can already be found in many cities, including Paris, Barcelona, and Milan (Tajani 2021). Cities to be lived on foot or by bicycle by leveraging the multifunctional services rooted to the territory and targeting the welfare and wellbeing of inhabitants. It is made possible by constructing solid networks of social and collective infrastructures, in which the system of activities and that of businesses function as social actors, moving beyond logics of mere economics and profit.

The neighbourhood is a physical and social space even independent of neighbourhood relations; places where it is pleasurable to live, where 'we feel at home away from home,' despite relations of life and work that also develop outside the neighbourhood. Everyone should have the possibility to access daily services in a reasonable period of time: a city made of neighbourhoods for living, dwelling and working, harmonising and minimising movements. A similar model of the city stimulates the diffusion of soft mobility and presents movement as a choice rather than an obligation.

2.1.2 Dwelling, Working, Living in Proximity

The theme of proximity involves the dimension of dwelling and services, and expands into fields of production and work. During the acute period of the pandemic, in both urban and suburban areas, work (in all of its forms) was parcelized and fractioned. It moved toward people and their homes. While it is plausible that the diminution in 'costs of distance' can generate phenomena of delocalisation outside large cities, the opportunity for remote work in broader contexts of dwelling, at a lesser cost, and with the ability to revitalise local histories and landscapes, cannot substitute urban social life and economics, though they can become an occasion for promoting frail and marginal territories. Working from home must be a choice and not an obligation imposed as a lifestyle, above all on more fragile classes (precarious workers, young people, and women).

Some research projects and experimental initiatives are working in this direction: the passage from home working to near working, programming subsidised spaces distributed across the territory and available to employees who can work remotely outside their homes: a form of neighbourhood working as an alternative to home working (Mariotti et al. 2021). We can redesign the geographies of work in a structural rather than transitory direction, beginning with the twofold criticality exposed by the pandemic: the majority of our homes are unsuitable at both the scale of individual units and entire buildings. This observation should direct our work in multiple directions: the redesign of housing, in its dimensions and toward a greater fluidity, reversibility/adaptability, flexibility, and 'open' multifunctional space, which can be dynamically configured thanks to devices that permit changing configurations based on changing uses and functions; the design of collective services and spaces in buildings for different types of use: recreational, cultural, for sport, work, both inside and outside buildings, intermediate semi-public spaces (or semi-private, depending on the point of observation), as filters between a block of flats and the rest of the city (courts, courtyards, condominium gardens, terraces, loggias, balconies, etc.); the location of offices based on proximity, as an alternative to concentrations in large headquarters.

For these reasons, the tertiarisation of the urban market, intended as the capacity to provide residential properties with integrated services (from community events and moments for conviviality to a condominium time bank), but also to manage building stock (owned and leased): the house is no longer considered property but a good, in some cases to be used for a limited period. Additionally, today, it is difficult to avoid the direct involvement of future users through guided forms of participatory design. For instance, by programming activities that can open up toward neighbourhoods and cities, focal points and spaces of aggregation for everyone, to unite innovation and social inclusion, increasing safety and, at the same time, the urban capacities of individuals and society. Moreover, favouring and promoting collective dwelling, founded on principles of relation and social integration. Like other cities, Milan has also recently tested innovative forms of *abitare solidale* (social housing): projects that favour the development of local communities, concentrating on individual flats and the urban block. These interventions focused on improving the quality of relations among diverse generations, promoting the active participation of residents and the transformation and utilisation of shared spaces, small shops, support to artisanal workshops, community spaces and the presence of allotment gardens. This experience fosters a new culture of social dwelling that favours a more inclusive society open to new lifestyles; and a profound and focused look at contexts and the evolution of society, economics, and culture.

With a possible new role of social entrepreneurship: the New EU Action Plan for the Social Economy (2021) promoted by the European Union places the economy and social entrepreneurship at the heart of the European Recovery Strategy. The idea is to attribute social enterprises a role in redistribution and transformative welfare, but also in industry and economic development, with a prevalence of social benefits over profit, the total or partial reinvestment of profits in activities of public or general interest and open and participatory systems of governance.

Proximity and Social Economy are positioned between industrial clusters at the foundations of plans to relaunch growth in Europe, considering the Social and Impact Economy within the perimeter of industrial policies for more equitable and inclusive growth. Under this profile, economies of proximity can represent true and proper ecosystems of innovation, with growth for businesses and technology, rediscovering the sense and role of social innovation; ecosystems capable of holding together economies and societies, extracting conditions of daily life from communities, building pertinent biographies of settled communities, founded on real local needs: a regenerative anti-fragility, not defined in abstract terms, but beginning with the needs of local territories and their specific historical, geographic and social conditions and building a gradual process of adaptation, aware that "(…) the best economy of cities is the care and culture of men" (Mumford 2002). Furthermore, the only care is a 'more just economy': social innovation and bottom-up transformation, individual entrepreneurship and the qualitative characteristics of local markets, all fundamental resources for managing the radical transformations ahead. The trust of communities, the value of participation, the inherent power that local economies and societies are the only antidotes to the poison of a culture that tends to cancel any dimension of complexity from the problems and tools necessary to solve them.

2.1.3 Milan, 'Mosaic-City'

Milan's history speaks to the construction of a city that incrementally absorbed different places and communities whose strong lasting identities configure the contemporary metropolis.

Despite its reduced administrative size (18,000 hectares), Milan is the de facto 'historic city' of a vast urban region. By exploiting the occasion offered by the pandemic to rethink our lifestyles and consequently our cities, we can imagine recomposing a contemporary organic mosaic, confirming this lengthy process of integrating articulated and differentiated identities.

In 1873 the Milan Municipality expanded its administrative borders by roughly three kilometres beyond the Spanish city walls, annexing the so-called *Comune dei Corpi Santi*, a name that derives from an Austrian health law that imposed the relocation of cemeteries outside the city walls, constituted in 1782 by Joseph II and articulated in six *sestieri* corresponding with the city gates (Orientale-Tosa, Romana-Vigentina, Ticinese-Lodovica, Vercellina-Castello, Comasina-Tenaglia, as well as the two gates controlling the city's canals, Darsena and Martesana), through which goods entered the city and where customs were levied. Annexed to the city, the villages and agricultural hamlets of the *Corpi Santi* became new quarters, while the farms be-came centres for the development largely of parishes, schools and important municipal buildings; they often lent their name to these developing streets and quarters.

This political process of aggregation was also promoted, at the territorial level, to confirm the imposing expansion of the manufacturing activities that soon charac-terised the city's economic growth: rural hamlets and villages, farms, parish churches and monasteries were absorbed by the urbanisation of the city: today some remain semi-central (Calvairate, Portello al Castello, San Pietro in Sala); others more periph-eral (Santa Maria alla Fontana, San Siro, Ghisolfa, Bovisa); others farther afield (Monluè, Barona, Gratosoglio, Tre Ronchetti).

A decree promulgated in 1808 annexed the *Corpi Santi* to Milan, together with all of the municipalities within a 4 mile radius of the city's bastions, that is, within 5 miles of the Piazza del Duomo. In 1923, as in other Italian cities (Naples and Genoa *in primis*), Milan continued the process of annexation-expansion with the aggregation of 11 neighbouring municipalities: Affori, Baggio, Chiaravalle Milanese, Crescen-zago, Gorla-Precotto, Greco Milanese, Lambrate, Musocco, Niguarda, Trenno, Vigentino, while Turro was aggregated in 1918. This new, extended ring included more economically and socially structured urban situations with respect to the first phase.

The process of aggregation that developed during the fifty years between the end of the 1800s and early 1900s saw the municipality's population grow to more than 800,000 inhabitants, reaching a total area of its current 18,000 hectares.

Milan's first master plans (Beruto Plan 1884–1889 and Pavia-Masera Plan 1913), which accompanied these processes and would guide subsequent urbanisations, essentially regulated only aspects of industrial functions and laid out the main street grid of the residential city, paying little attention to values of history-identity, such that numerous intellectuals from this period expressed their concern over the effects of such a strong expansion of building, and the risk of losing a consistent historic-artistic heritage: in 1924, the *Società Storica Lombarda* formed a committee of experts to survey historical and artistic heritage in the aggregated municipality, to promote its preservation.

The anniversaries of these two aggregations offered the occasion for a historical–critical reflection on the foundations of such a significant part of Milan's urban history, which stimulated the preparation of an appeal by a group of scholars and intellectuals (January 2022). Among other things, they proposed the creation of a network and the organised and institutionalised promotion of an awareness of the city's urban history, with particular reference to the processes of aggregation, to better understand the heritage of the Milan's different communities, respect them and, above all, care for them. Milan certainly demonstrates an intrinsic ability to continually change and innovate and face the challenge of dealing with the material and cultural traces of its history, capturing its values and potentialities.

Milan is subdivided into nine *Municipi* (boroughs), geographic and administrative units that vary from just under 100,000 to just under 190,000 inhabitants: de facto 'cities in the city,' each characterised by its own geographic, historic, social and cultural identity and revealing a constitutionally polycentric city. The *Municipi* are in turn articulated in 88 NIL (*Nuclei di Identità Locale*, Nuclei of Local Identity), the city's true and proper neighbourhoods, with physical, identarian and symbolic confines (Galster 2001) in which it is possible to recognise historical and contemporary settlements, each with its own specific characteristics. The NIL, introduced by the PGT 2012 (*Piano di Governo del Territorio*, City Plan), represent proper systems of urban vitality, concentrations of local commercial activities, landscaped areas, spaces of aggregation, services and facilities; existing and/or to be improved and designed, through which to organise an articulated system of local and urban facilities; a true territorial atlas, a tool for verifying and consulting programming services, expressions of territorial phenomena representative of local dynamicity. They are dense toponyms, stores of memories, many with the characteristics of true and proper historical-social villages (Affori, San Siro, Barona, Gratosoglio, Ghisolfa, Bovisa, Calvairate, Monluè). Milan is a city that has rediscovered its identity as a sum of different parts. This is also true of its cultural offering, which takes the form of a strategic plan for museums (*Una città, 20 musei: 4 distretti. Verso un piano strategico per i musei di Milano*, "One City, 20 Museums: 4 Districts. Toward a Strategic Plan for Milan's Museums", promoted by the City of Milan's Directorate-General for Culture in 2019). This programme reinvents the relationship between civic museums, the city and public fruition, connecting the offering of culture with urban, economic and social development strategies.

Milan is a 'mosaic-city' that can be imagined differently. This process can begin with the tiles represented by its neighbourhoods (NIL), each capable of offering essential services based on proximity measured in distances that can be travelled on foot or by bicycle: home, work, local shops and services, schools, health and cultural facilities. The idea is to create community hubs, rethink commerce (Tamini

2020), neighbourhood parks (Nucci 2012), the territorial healthcare system defined by the *Case della Salute* (outpatient centres)[1] (Giarelli and Vicarelli 2020) and the neighbourhood school system (Mattioli et al. 2020) as indispensable outposts and urban landmarks for a new collective existence. Urban mobility represents another strategy for structuring public space (Campioli 2020) by introducing innovative plans that combine living, working, care, leisure and social interaction (Karsten 2003) to provide access to places of work, accommodations, food, healthcare, education, culture and free time (Moreno 2020).

2.1.4   Neighbourhoods, Community and Public Space

The proximity dimension typical of the neighbourhood imposes a coordinated design of housing, spaces for work, leisure, commerce and services. The theme of public services and space is the expression of the sociality and vivacity of a city. Therefore, in any city plan, it is indispensable to initiate a process of urban regeneration that occurs precisely through the neighbourhood and its public spaces. A process that must start with site-specific projects that care for every day (the remedy is the cure), searching for a new urbanity (access to healthcare, housing, work, culture and free time) which becomes a generator of new urban experiences. In synthesis, we are searching for an effective and safe relationship between the shape of urban space and social life. This is coupled with a growing awareness of place made possible by forms of choral production and social projects. These are the elements of a system to be reimagined and redesigned within a new and contemporary common ground, represented at street level by open and permeable buildings, spaces of encounter, exchange, and interaction, offering new typologies of public spaces, collective uses, activities, and places of aggregation (Gehl 2012).

The reference is to authentic and proper interactions with the city, animating the 'common ground' of the quarter, which becomes a space of encounter, exchange, interaction and urban permeability. Each place speaks of different forms of encounter and the attraction of flows of people, guided by the desire to participate in necessary and/or voluntary activities (Gehl 2011).

The proximity network is a system of open spaces, both green and mineral, which will permit a renewed continuity and new typologies of spaces and uses. It includes porous lots, thanks also to vegetation in courts and courtyards, which will make it possible to define a new system of urban relations among private and public spaces, to favour their crossing.

The zero level of the city is the space of co-penetration between different activities and uses, not to mention the fundamental prerequisite for guaranteeing an

---

[1] The *Case della Salute* (outpatient centres) are a unique point of reference where citizens can access primary health care services through the collaboration between family physicians, specialists, nurses and healthcare workers.

urban *mixité*; new accessibility, made of juxtapositions and overlappings of different networks, needs and practices.

This space is home to settlements, where the 'feet' of buildings are occupied by 'proximity economies' (small shops, local craftsmen, public establishments), whose importance has recently been re-discovered: the baker, the greengrocer, the grocer, the newsstand, which once animated neighbourhood life and once appeared activities of little use to our contemporary era of consumerism and metropolitan lifestyles.

Learning from the historic city, this contemporary urbanity mixes different activities and functions, which are deeply integrated and engaged in a dialogue with the open space system: a constellation of uses and activities to be promoted, as it was with the first City Plan (PGT) from 2012 for Milan (Arcidiacono et al. 2013), which recognised the social and collective value, the role of vitality and at the same time as an urban garrison.

Within this framework, the space between objects assumes the same importance as the objects themselves. These spaces welcome practices; they are spaces to be designed and in which to reflect on inhabitability. The goal is to dilate public and collective space until it forms new networks of shared and accessible areas capable of improving the quality of dwelling and developing a sense among residents of belonging to a particular part of the city. The idea is to re-configure a sequence of open public and private spaces that are flexible and accepting of diverse uses and offer opportunities for new activities and practices of daily life.

Proximity is reimagined as richness, a social and economic value, this is the possibility generated by the health crisis: the opportunity to radically reform our homes (our refuge, intimate and domestic space), that had been flattened to meet market standards wholly indifferent to changing conditions and lifestyles; and our cities (the areas in which we dwell), our way of living in the world.

The objective is to construct a settlement system articulated as an organic and integrated archipelago of quarters with strong relations, privileging public mobility and pedestrian and bicycle connections. This system comprises a collection of neighbourhoods with their own identity within a simple urban composition. Creating articulated urban centralities, true epicentres/focal points for neighbourhood life, outlines the ambitious objective of constructing a sequence of urban space and distributed polarities. Not an ordinary urban fabric, but genuine pieces of a mosaic with their own identity; in this manner, promoting the services and values of proximity also rediscovered due to the health crisis induced by the pandemic.

Each project raises a theme of identity. Building an identity is both an opportunity and a process that must seek the most virtuous correspondence between the characters of place and functional vocations. Guiding and orienting transformations by defining and emphasising existing 'lines of force; implementing a strategy for adapting to context often ignored by modernist design; working patiently with its elements to build solutions founded on and motivated by learning about and interpreting it. Collective space thus becomes a 'generator,' a binder in a sequence of different spaces, whose articulation is closely linked to the fabrics that define its edges that identify a pattern capable of giving meaning to open space.

The contemporary city must also be reimagined and redesigned so that people, moving on foot or by bicycle, can experience the essence of urban life. This is one way to recover the relationship with time cancelled by modernity through its separation from space and the rigidly functional deformation of its meaning. The times and rhythms of the city have been the object of urban studies, policies, and actions during the twentieth century in diverse Italian cities. Experiments focused on defining the movement of public action from public hours to social and urban times, to the design of the city, to time-oriented urbanism (Bonfiglioli 1990; Mareggi 2011). These experiments begin with past reflections and an awareness of the current complexity of the space–time relationship and rhythms of the city. Moreover, they begin with the different populations that inhabit the city (Pasqui 2008) and the consequent necessity to design a system of hours offering related services according to a logic of diversification and de-synchronisation. In addition, some policies and actions demand a necessary and strong direction from the public sector (Zajczyk 2007). Proximity is intended as urban distances and spaces at the human scale, capable of redesigning urban and territorial hierarchies. The rediscovery of the neighbourhood and the scale of proximity can redefine a new articulation of space–time, cared for by communities. In fact, social resilience and urban regeneration begin with a new idea of proximity and policies of caring for every day spaces and objects (Manzini 2018).

2.2 Resilient Urban Structures

2.2.1 Urban Transformations as Missed Opportunities for Regeneration

Existing cities are transformed slowly. They are driven by an extensive, spontaneous, pervasive and diffuse metabolism, made of transformations bound more to function than to questions of morphology and settlement. We generally understand them only after transformations have been completed, as these processes unfold gradually, revealing themselves only when the face of various neighbourhoods has already changed (Pareglio and Vitillo 2013).

The important urban transformations in Milan, which reutilised the spaces of abandoned factories, in the wake of changes to the manufacturing sector and the characteristics of the city's history, from the 1990s (with the introduction of particular planning tools such as the *Programmi di Riqualificazione Urbana*, PRU and the *Programmi Integrati di Intervento*, PII) to the preparation of the first City Plan (PGT 2012), were large-scale projects, specific, discontinuous, and disconnected from an overall vision (Oliva 2002; Arcidiacono and Pogliani 2011). In reality, these large projects overshadowed the extensive transformation of the ordinary city, consistently lacking an organic project for its care, not to speak of its regeneration.

This difference in attention is also manifest in other cities. However, it is more evident in Milan, owing to a particular overlap of interests that have made urban land value a tool at the service of financial gain (Agnoletti and Di Maio 2011; De Carli 2011; Gallino 2011; Sapelli 2011). This has conditioned the relationship between public–private and the distribution of urban functions and, more generally, the characters, functions and values of the urban market (Curti 2007). An example of "relational capitalism," if not crony capitalism (Stiglitz 2002; Krugman 2008), is when business fails to live up to its natural condition of being an assumer of risk.

A context in which not even the so highly sought after and much-touted concept of 'flexibility' has functioned, as demonstrated by the lengthy times required to develop and build projects, despite the use of negotiated planning instruments, particularly the PRU and PII. Similarly, the 'social redistribution' of land value also failed to function, or functioned poorly, as proven by the modest advantages to the public sector generated by transformations: in the best European examples, a significant quota of land value is instead acquired by society (Camagni 2012; Camagni and Modigliani 2013), through a balanced exchange between private and public benefits, based on measurable and comparable dimensions.

In Milan, the interests of the building and real estate sectors, represented by large urban projects, were aprioristically indicated as drivers of economic growth, based on a now-obsolete model (Palermo 2011). On the one hand, this approach ignored the quality of individual projects and, on the other hand, overshadowed the needs of caring for and regenerating the existing city that, instead, could have been an opportunity for activating a diverse model of urban development.

However, the ordinary city was transformed, driven by a spontaneous, pervasive and diffuse metabolism, made of reparations and transformations that were more a question of function than of morphology or settlement, and which revealed principal moments of reuse: decommissioned manufacturing districts and mature urban environments. This metabolism revealed critical aspects where it was not grafted onto robust and complete urban structures, or when operations of substitution dealt with older parts of the city, such as decommissioned industrial and craft-based areas located or distributed in the city. These criticalities were understood only after the transformations had been made; because they proceeded incrementally, by degrees, almost exclusively based on building permits or simple communications, meaning they showed their face only after the appearance of many neighbourhoods had already changed.

The reuse of decommissioned manufacturing areas occurred principally through interventions of building renovation—whose notoriety is largely attributable to the phenomenon of loft living—in the city's first periphery, for the most part along the edge of the railway, in particular in the industrial areas laid out in the 1976 City Plan (called *Piano Regolatore Generale*, PRG). In the majority of cases, this process involved medium to small-sized areas (from 1 to 3 hectares), resulting from the process of deindustrialisation of the manufacturing sector that occurred in the city from the late 1970s (La Varra 2007), which also led to the construction of new residential villages, whose effects reverberated throughout neighbouring urban fabrics.

In general, they were transformations to the functions of important parts of the city, accompanied by modest modifications to buildings, which generated new typologies of settlement and dwelling: from abandoned factories to new forms of dwelling, to new small residential quarters, to the arrival of new activities. More often than not, like parasites, utilising the public and private services present in their immediate surroundings. This is true even where transformations were the result of negotiated plans and programmes, and even in the case of specific transformations, at the limits of urban rules and regulations, and in some cases beyond them.

Under a real or apparent urban vivacity, the vibrant appearance of the phenomena of the *movida*, of informality and creativity, three unresolved questions remain: mobility and the environment (above all for soil remediation), primary urbanisation works (above all for energy and technology networks), primary and immediately accessible services (above all parking, but also neighbourhood shops). Another issue in need of resolution involves the opacity of the market: it is well-known in Italy that properties classified in cadastral category C3 (laboratories for arts and crafts) are utilised as residential properties in Milan, from Via Mecenate to the Bicocca, from Via Ripamonti to Via Meda. Without the fascination of Tribeca or Greenwich Village, they have allowed their owners to save from 15–25% on the purchase price (OSMI 2012; CCIAA Milano 2010).

2.2.2 Nineteenth Century Urban Fabrics

The urban structures of the nineteenth century have shown themselves capable of adapting to phenomena of reuse, manifesting ductility and resistance to these processes, concerning typologies and forms of settlement and the offering of services. In particular, Milan's nineteenth-century urban fabric proved capable of clearly expressing defined, educated and gentle forms, easily recognisable as familiar and civic; forms capable of providing us with a sense of harmonious relationship with others and the spaces we inhabit (Tonon 2014). It is a fascinating condition in which we could define a capacity to adapt to pressures and modifications (functional, typological, morphological, and even performance-based). We could accompany Berlage's Amsterdam, or Cerdà's Barcelona (1884–1889) with Beruto's Milan (1884–1889): a 'cautious and modest' project, spoken in undertones and dressed 'in plain clothes' (Oliva 2002), with weak prescriptions in terms of land use and a robust urban grid, articulated and constructed to last over time, and with clearly recognisable measures and canons of development (few and simple).

The Beruto Plan treated urban form using rules for architecture and morphology, avoiding the regulation of functions, except for large public and industrial uses (sometimes). The result is a non-ordinary and sophisticated beauty of the city, which has proven capable of resisting even some impertinent interventions of architectural substitution, absorbed by the design, quality, and composure of urban space and scale expressed not in single buildings but in the choral nature of virtuous and synergetic relations between public and private space. It is a sum of non-specialised and programmatically ordinary urban and architectural spaces that do not age but adapt to a contemporary era of new urban uses. Milan is an ordinary, civil city that has adapted to a fluid conversion to new functions thanks to a typological configuration which naturally includes modifications to uses.

In synthesis, the canons of the Milanese urban block, whose design is a sum of dimensions, stereometrics, building typologies and relations between public and private spaces, have demonstrated a natural resilience, an inherent strategy of adaption to change. They are the expression of a society that remains capable of adopting an integrated approach to different ways of dwelling. The Milanese block has managed to provide an ordinary quality with a minor soul, a form of sober and simultaneously resistant beauty freed up in the folds of the consolidated city.

The specific reference here is to the majority of areas listed in the PGT 2012 as "Areas with a Recognisable Urban Pattern/Morphology" (*Ambiti a Disegno urbano Riconoscibile*, ADR), in particular those located within the outermost ring road. They are prevalently urban fabrics generated by master plans from the late nineteenth and early twentieth centuries (Beruto Plan 1884–1889 and Pavia-Masera Plan 1912), which proposed a unified urban plan, rules of development and an architectural language, with an implicit strategy for adapting to change, integrating and building system of settlement together with structure of public space.

When planning resilient urban structures, a fundamental role can be played, finally, by green and blue networks, as better described in Chap. 4 of this book. These multifunctional and interconnected platforms can play a decisive role in producing environmental, ecological, climatic, and social benefits for the regeneration of the contemporary city; to be imagined, designed and built at diverse scales (city, quarter, block, building), to promote and develop Natural Capital (Schumacher 1973) and improve the overall quality of life in the city.

2.3   (Re)shape Spaces. From Housing to Living

2.3.1   Looking for Urban Quality

Nowadays, urbanism is almost exclusively the day-to-day management of the existing city, which expresses a renewed attitude toward caring for places, intended as the capacity to interpret contexts. It moves toward a model of a more liveable, resilient and technologically advanced city that promotes new environmental models of prevention, mitigation, and adaptation to risk and climate change. Despite this, space possesses an intrinsic hardness and resistance to transformation: the physical matter that shapes it naturally resists attempts to (re)model it. The regeneration of existing heritage is indispensable to improving inhabitability. This is particularly true when adapting the dimension of housing units to the new characteristics of contemporary families and lifestyles; learning to work incrementally with the existing, re-stitching, mending, reusing, similar to ancient artisans who adjust things with care and patience.

This demands we work with three seemingly crucial themes: homes and cities that lack quality, the banalisation of commercial activities, and waste (Vitillo 2011).

Homes and cities that lack quality. For far too long we have been building homes according to obsolete standards and functions, for men, women and family nuclei that no longer exist and perhaps never existed. We no longer apply research to architectural types and dwelling models concerning new populations, new typologies and morphologies of societies and families to new needs of social-assisted living. We have forgotten the lessons and investigations that explored and articulated the typology of the home through in-depth study, attentive and focused on different contexts and the evolution of society, economics and culture. At the same time, we have forgotten the urban morphology, its elementary rules, studies of settlement models and the values of urbanity that characterised Italian cities. Most recent programmes and projects are indifferent to context, a-topical and self-referential. They fail to reflect on the relationship between built and open spaces, how buildings touch the ground and functions that serve as a hinge between private and public space, rigidly separating areas according to zoning requirements. The result is a loss in the necessary integration between urban morphology and community values, the construction of the urban fabric as a social, economic and cultural value, and the result and material expression of its liveability. This occurs despite a tradition that set standards for elegant, intelligent and sophisticated interpretations of urbanity. In this manner, we waste one of the most beautiful legacies inherited from modern Italian architectural culture and projects that managed to innovatively interpret neighbourhoods as spaces of social existence.

The banalisation of commercial activities. We have substituted dwelling needs with a parody and reality made of low quality, artificially sweetened and banal surrogates. One of the problems of globalisation and mass society is the perceivable nexus between poor quality and commercial success, in a sort of absolute dominion of mediocrity. In the era of globalisation, of homologation and commoditization, we have passed without interruption from the idealism of 'inhabitable space' to the barkers of televised commercials, who have set the tones of markets and trends much more than we are willing to believe.

Waste. We have produced two forms of waste: on the one hand, the degradation of the territory that has come to characterise the contemporary city through inefficient sprawl (land take, pollution, instabilities, decay and divisions), with unsustainable infrastructural, environmental, economic and social costs. On the other hand, the waste of buildings, represented by a housing stock comprised of units much larger than the European average, topologically obsolete and, what is more, constructed using energy-absorbing materials and techniques and very expensive to maintain. Today, urban design has lost its capability to reflect reality, whose construction and management require a skill and the recovery of the values and qualities of the historical city, which allow it to be perceived naturally, as if it were our home (Zucchi 2021) based on three indicators: scale, non-specialisation and imageability.

Scale. In many cities in recent years the distance between the human body and its home and between the human body and the city through which it moves have been progressively dilated and extended. This is true of both dimensions and relations, to the point that any measure at the human scale was progressively ignored and finally abandoned (Anders 1963). The bond, relations of continuity, proximity, the values that built the urban fabrics of the city inherited from the past, by analogy with the human body but also with the social body, have been weakened over time, such that today the distance between buildings has become a dominant, if not exclusive, criterion (health-sanitary regulations).

Non-specialisation. The city is a world in which everything is mixed: uses and activities, wealth and poverty, ethnic, political, cultural, and religious diversities; a true and proper dynamic and transformative 'mixing machine' whose amalgam enlivens the urban fabric, creates exchanges, shared moments and human relations. The recognition of value and quality we attribute to the historical city is also born of its strong functional, social and morphological integration. One of the central themes of urban regeneration thus becomes the search for *mixité*, both functional and social, combining residential uses with others related to work, shopping, free time, and contrasting the mono-functionality characteristic of urban peripheries across Italy. Not a traditional (services, housing, activities) but an innovative mix that includes free time, sport, culture and entertainment. Integrating functions within buildings and across the entire city (multiuse city) (Cacciari 2004; Coupland 1996); and contrasting modernist hyper-functionalisation and the functionalisation of dwelling proposed by homologising and trivialising real estate developments.

Imageability. The recovery of the human scale and measure makes it possible to free up another important urban capacitation (Alessandrini 2019; Nussbaum and Sen 1993): imageability, intended as the clear recognisability of physical space, the legibility of its configuration, of an architectural and urban syntax; the perception of its dimensions, of its relations, of the places of which it is comprised, of the activities that take place within it, and the restoration of the recognisability of place.

2.3.2   Urban Regeneration, a Multi-scale and Multi-dimensional Project

Urban regeneration, a concept that has spread in relatively recent times, quickly took hold and has become inflated to the point of becoming a 'blanket' term. This passe-partout should be utilised with greater pertinence, parsimony and attention. We can reasonably affirm that urban regeneration is substantially different from what we referred to for years as urban requalification. While this latter essentially took the form of a discipline-specific project, regeneration is a multi-scale project, a systemic challenge that holds together a plurality of dimensions: of settlement, but also economic, social, related to energy, the environment, landscapes, institutions, and participation; manifest through both integrated public policies and through spontaneous, ungoverned processes.

Public policies in Europe are often constructed with a multidimensional character, oriented toward redefining rules and forms of welfare, both local and material; to combat exclusion by increasing opportunities for social groups; constructing conditions for local economic development. The regeneration of brownfields in many European cities often pursues solutions that bring overall environmental effects while guaranteeing new and qualified urban offerings, utilising concrete approaches to design. Regeneration can also be a spontaneous process, neither planned nor designed: a silent urban metabolism, not easily perceived in the medium-long term, which changes crucial pieces of the city, almost without us being aware. This is a process of micro-transformations, not governed by the tools and regulations of urban planning but led by multiple and articulated socio-economic desires and by simple buildings.

We can observe both approaches to regeneration in Milan, from the programmatic to the planned. Public land plays a leading role in each. It offers occasions for new economic and social infrastructural projects (the ex-rail lands, military barracks, the former Expo site); or occasions with ordinary and spontaneous characteristics that regenerate obsolescent building stock.

Granular urban regeneration—of single buildings, complexes of buildings, entire city blocks—like processes of reuse and a functional metabolism extended over time and space, can be programmed, promoted and guided (Lavorato et al. 2020). Naturally, all processes and transformations involving a rich and articulated historical, cultural and material palimpsest, such as that represented by European cities, target a long-term that, however, we can and perhaps must direct from the outset by exploiting 'the window of opportunity' opened by the Covid-19 pandemic. In particular, this emergency has taught us that we must be capable of constructing programmes and devices that 'make room,' that seek a certain 'redundancy' in the system of open and green spaces (public and private). However, the same also applies to those areas and facilities necessary for the realisation of social housing, in all of its different and multiple articulations (from public housing to inclusive rentals, to subsidised rental and sales). Inevitably, this process must consider the physical dimension of places, with environments that present the least possible specialisation and are hopefully anti-fragile.

Contemporary cities are living organisms, a mass of networks built by the unpredictable actions of countless and chaotic actors (Amin, Thrift 2016). To regenerate them, we must be skilled in building outside the pleasing and self-referential academic abstractness that pervades professional debate. We must use convincing and straightforward devices for regeneration, capable of unfolding over time. We must adapt to circumstances and exploit available resources, moving along different levels to build frames, rules, projects and actions within a process that does not pursue a single and pre-established objective, but is redefined in practice (Gabellini 2010; Galuzzi 2010). There is a need for a framework armature for grafting specific interventions (resilient pieces), rooted in contexts and process-oriented and adaptive (Galuzzi and Vitillo 2021). Interventions participate in its realisation and their destiny can also mature over time with changing social and physical structures. In this perspective, the ecological and environmental network can represent the dominant spatial figure, capable of exploiting and giving form and quality to the contemporary city. With stubbornness and patience we build a culturally and professionally founded attitude toward design (Galuzzi et al. 2018). An approach capable of building relations among the multiple dimensions characteristic of urban regeneration, and translating them into actions integrated at the most appropriate scales of intervention through a strategy involving multiple scales and multiple actors (Bevilacqua et al. 2020). A strategy capable of welcoming the ability for urban communities to create and capitalise on social value, using mechanisms associated with a proper production of services and functions for the community, based on networks and social entrepreneurship (Calderini and Gerli 2020; Alessandrini 2019). A strategy for transformative and adaptive regeneration that works with a set of constant characteristics, such as modularity, control of scale, reversibility, sobriety, temporary use, contrasting the private appropriation of land value, equity, the development of skills, guaranteeing the right to the city (Galuzzi et al. 2019). Curing, managing, evaluating, more than predicting and dimensioning, on the productive and active role of ecosystemic services (Arcidiacono et al. 2018), founded on principles of protection and promotion, but also on processes of transformation (Oliva and Ricci 2017), capable of integrating social, economic and urban development.

In synthesis, a transformative and generative regeneration, able to favour an enabling context and a cohesive ecosystem of actors capable of co-producing, with the community, places that serve the wellbeing of citizens (Galuzzi et al. 2020), promoting "relational flows" and the "pleasures and rituals of collaboration" (Sennett 2012). A similar regeneration enables measures to prevent change and welcome modifications over time. Regeneration favours the capacity to interpret fragile contexts and, at the same time, to mediate the construction of an expanded and bottom-up approach to design (Terracciano 2014). By paying attention to dwelling, a set of ordinary and widespread actions and extraordinary and intensive actions, sustainable development practices are oriented toward favouring good behaviour and penalising improper practices (Thaler and Sunstein 2008).

Many claim—and with good reason—that after the pandemic everything will return as it was before: however, despite everything we have certainly rediscovered new activism and a strong centrality of public policies, and regulating actions; this is due not only to the vast resources of the National Recovery and Resilience Plan (*Piano Nazionale di Ripresa e Resilienza*, PNRR 2021) that, in any case, we must not forget, will weigh on the shoulders of future generations. Almost all reflections propose a new and radical paradigm and alternative development model, able to respond to the ecological-environmental challenge, climate change and the multiple risks faced by societies and territories.

There is a rich and exciting debate around a true social crisis with significant impacts on healthcare (Saraceno 2021). A reflection characterised by proposals and solutions that have suggested different methods of discontinuity, to be explored so that we can learn from the lesson of Covid-19.

Therefore, as useful orientation tools, we can attempt to draw an initial map, necessarily rough and still out of focus, of the issues the city must explore in the near future. There are two preliminary assumptions and a few guidelines, which presuppose a change, reformist in its methods and radical in its content, of the paradigms of the plan and the urban project. First and foremost, the two preliminary assumptions: the importance of space and the ineradicable dimension of risk.

Space Matters. A reformist attitude is founded on the necessary awareness of how much and to what degree space, structure and organisation count as fundamental and unavoidable physical elements. In the urban and territorial economy, space has reacquired a central role, that of 'making space,' with a return to Keynesian policies, to an organic notion of the territory.

Digital networks, technologies and infrastructures, which we often naively consider solely in their immaterial and incorporeal dimensions, function through the significant modification of space and its physical characteristics.

Incorporating the Dimension of Risk. In an informed manner, we must assume risk as a structural component of society, the city, and territories (Beck 1992); it must be placed within a multi-dimensional perspective: climate, environment, health, economy-finance, terrorism. This means learning to coexist with risks and incorporating them within our forecasts and projects. Climate change and associated risks are no longer the stuff of abstract academic research or episodic events far from our everyday lives; they are now common events that affect the present in which all of us live (Vitillo 2022).

A plastic and visual representation of how these changes affect our daily lives can be found in the elegant and sophisticated physical-geographical maps in the book by (Pievani and Varotto 2021), documenting our lengthy voyage through the Anthropocene: a visionary geography of our future, not necessarily the most probable, but certainly plausible, which imagines Italy one thousand years after Goethe's Italian Grand Tour when extreme events and the complete melting of the polar ice caps will have raised the level of *mare nostrum* by 65 m (Pievani and Varotto 2021).

2.3.3 How to Bring About Regeneration?

It appears possible to identify a few operative guidelines that, as mentioned, presuppose a shift in design paradigms, to once again say something concrete about, and for, the reality of people's lives (Solero and Vitillo 2021).

Rediscovering the Values of Proximity. This rediscovery at a time when physical and social distancing is a rule we must necessarily respect, may appear out of time and out of play. In reality, cities are what make it possible to weave unique social, cultural and economic relations between people, which is an inherent characteristic of the urban world. Cities are based on the social interaction permitted by proximity, neighbourly contact, face-to-face experience (a proper technology of communication), on the discovery of the unexpected, allowing for an exchange of know-how that cannot be codified or transmitted remotely.

Direct and physical encounters stimulate active participation and imply mutual commitment. There is no collective growth without the authentic sense of sociality, another reason why cities are not destined to die.

How can we bring this about? In a contemporary key, re-imagining the value of relations of proximity, which form the urban fabrics of the European city inherited from the past and that the world envies us. This awareness gives birth to building the city as an 'archipelago of quarters,' to be lived on foot or by bicycle, currently being tested in many cities, including Paris, Barcelona, and Milan. Leveraging a renewed multifunctionality of services and their rooting in the territory, aiming toward the welfare and wellbeing of its inhabitants; constructing solid networks of social and collective infrastructures, in which the system of activities and enterprises function as social actors, overcoming the base logic of economics and profit.

Within this perspective, the design of open spaces, inherently trans-scalar, intersects multiple scales that any talented designer must be able to navigate.

In particular, the system of landscaping that, as we learned some time ago from ecological-environmental professions, must be designed as an interconnected network of points, lines, and surfaces to be linked at different scales: from the extended networks of large green systems (suburban and urban); to the intermediate networks of neighbourhood parks; from the networks of green courts of proximity (at the scale of the block/quarter), configured as true infrastructures. Green space is not only a number, a surface, a function, but an integrated and complex device: ecological, landscape-environment, but also social, healthy and for safety. A system of open spaces and green spaces configured in the form of networks, that serves as a fabric of ecological regeneration and for improving the hygienic-sanitary conditions of the urban ecosystem, contributing to the design and structure of its form, organising and activating relations and connections among its parts (Angrilli 2002).

Working with humility, abandoning manias and presumptions of control, often pursued through a detailed and hyper-deterministic project, which has characterised the entirety of architectural culture as well as the design of landscaping during the modern era.

An effective and pertinent design of green space should probably begin with the configuration of a very simple temporary use: a planted lawn, to be watched over by a patient eye and cared in its use and adoption by its users: they are the ones, with their behaviour and attitudes, who determine its evolution and identify its characters.

A necessarily temporary and incremental project, easily reversible, constructed over time based on its effective utilisation by different urban populations.

Setting Out From the Structural Dimension of Problems. We must necessarily modify diverse themes managed by those professions that work with space, setting out from the structural dimension of problems.

Density, the system of mobility and open spaces, urban welfare, the central hinges of modern urbanism and contemporary dwelling, are spatial categories that the pandemic asks us to reconsider, with the rediscovery of the importance of the themes of relational proximity, of both our bodies and of urban space.

This begins with incisive care for social and spatial inequalities, for social and cultural capital gaps. We have witnessed a frightening increase in inequalities, which can be read as a watermark even when we approach the theme of housing and dwelling, which has compromised the balance between citizens, governments, states and private businesses (Ciocca 2021; Coppola et al. 2021). We live in a time when the power of some large multinational corporations surpasses that of state governments in a series of fields traditionally reserved for them: this transformation, also read in its historical development, probably necessitates the rewriting of a social pact (Ross 2021). A document that must confront three concepts in particular: the limit of development (not to be overcome), defining strategies for avoiding irreversible crises and degradations of society and the environment; social inequalities, contrasting the macroscopic disparities that characterise contemporary society (a true obstacle to development and perspectives of growth); settled communities (true garrisons of places, to be cared for).

Enabling the Future. For designers, the problem of the future is essentially a question of society's desire and shared visions, which are possible, pragmatic and realistic.

Under conditions of uncertainty, it would seem indispensable to activate non-predictive decision-making processes: a framework regulation, with margins of movement, but capable of ensuring the conditions that permit reconsideration of how the city and its articulated metabolisms truly work (function and settlement). Setting conditions, orienting, without predefining: until society can freely express its social and economic behaviour, its own and autonomous trajectories of life. Against any rigidification of design and against any temptation to define a definitive urban landscape; on the contrary, leaving time for articulation, for complexity and the rich-ness of processes. After the lengthy season of real estate-driven operations we must work to utilise devices integrated with history, culture and materiality of cities and territories. We must orient choices in light of the possibilities for action and by caring for fragilities, imagining what is possible and opening up ideas toward the future, seeking to hold together the efficacy, breadth and articulation of possible scenarios. This collaborative and inclusive process permits the construction of an idea of the future, helpful in coordinating our actions. The time for assuming stable models and making long-term forecasts has passed. We must now program by scenarios, intended as true systems, open to all possibilities. Scenarios permit a community to learn and become responsible, which, at the same time, favours the "capability approach" (Nussbaum 2011; Alessandrini 2019). The future is not an unknown, but a resource not to be wasted (Ferraro 1998); two beautiful considerations on the future, expressed perhaps not accidentally by a writer and a poet, respectively, help put things in perspective: "As for the future, your task is not to foresee it, but to enable it" (Saint-Exupéry 1951); and "The future enters into us, to transform itself in us, long before it happens." (Rilke 1980).

Programming Redundancy and Anti-Fragility. Urban space has a social quality because it is a source of human relations and interactions. During the pandemic, the images of empty spaces in our cities exposed the 'fourth dimension' of space, its cultural nature and the current importance of 'social space.' These are phys-ical spaces emptied by the health emergency, which we have already returned to spending time in and sharing, appreciating physicality and contact and reconsid-ering their fundamental physical-spatial dimension and innate characters and prac-tices. However, programming a certain redundancy, and their virtuous anti-fragility, is intended as a precautionary principle. Space is intended as a condition of possibility and a constituent factor of action and a concrete and physical-corporeal existence in the world, experienced to the fullest, in its luminous Baudelarian definition. We assume spatial thinking as the privileged way to access the concrete forms of life and the action of different subjects and actors.

We must redesign physical and social spaces to be temporary, non-specialised, polyfunctional, hybrid and reversible. The new design perspective is a reservoir of skills and abilities that develops adaptive processes to generate responses appropriate to unexpected situations—made increasingly more possible, probable and frequent by the growth of complexity—abandoning elitist, melancholy, aestheticising and nostalgic visions that praise low-density and the flight from the city as a possible response to the pandemic.

On the contrary, we must reconsider the forms and practices of urban life in operative terms, overcoming the false dualism between urban-centric and localist-decentric visions, as well as the opposition between the beauty and simplicity of life in small villages and the chaos of urban risks. Prefiguring a comprehensive and integrated project for territories, rethinking contemporary cities and territories together; triggering an ecological transition (the green shift) toward models of development without growth centred on the quality of dwelling.

In particular, through balanced processes of urban transition, beginning with the role of landscaped and natural space in cities, structured through the definition of frames at multiple scales: green networks and nodes, from the territorial scale to the urban scale, to the scale of the quarter to the scale/dimension of proximity; processes that must inevitably be multi-level, multi-actor, multi-sector.

The design of the environment, open spaces and ecological networks represents the primary key to interpreting current urban and spatial planning policies. Open spaces must move from being a "design material" to a "structuring element" of metropolitan landscapes.

Architecture and urbanism do not end with their respective programmes of functions; they must behave like a soup bowl, functioning as true containers for our liquid lives (Zucchi 2014). An articulated and composite set of practical know-how to be tested with the conditions of life that make it possible. More than generating new lifestyles, they must be able to host them and amplify them: life is inherently disorderly, unpredictable, and confused; architecture and urbanism must know how to tolerate and interpret it.

The design of public and open space should behave in the same manner: responding to the specific needs and questions expressed by settled communities while also being capable of welcoming variations and differences, the dimension of the unexpected and the un-expectable. With clarity of form that must be reinvented each time, without becoming nostalgic, but exercised with a principle of spontaneity capable of releasing a sense of naturalness, a beauty that is only apparently natural and casual: borrowing a term from the art of gardens, we could define it as *sharawadgi*, the naturalness of the world opposed to rational construction.

The field of interventions must therefore be expanded to include relational aspects and links with the consolidated city: looking not only at housing, but also at the quality of dwelling as a broader and more complex theme; testing housing solutions that differ greatly from the trivialising models offered by the real estate market; reflecting on the relation between built and open spaces, on how buildings connect with the ground, on uses that serve as a hinge between private and public space. Similar projects have managed to innovatively interpret neighbourhoods as spaces of collective life, searching for the (spatial and social) affordability of context. They have favoured a social mix, representing one of the unique and founding values of plural and inclusive cities and civil and collective life. If we were to suppose that affordability is more than just housing, and that programmes can no longer be sector-specific, the passage form public housing policies (*politiche per la casa*) to social housing policies (*politiche per l'abitare*) would seem inevitable.

The question of design consequentially becomes increasingly more one of context: the attractiveness of cities, intended as the capacity to attract new urban populations, is not mechanically and simply linked to the quality of what is built, nor does it depend necessarily on income levels, as was once the case. A successful real estate and development project is now one that best agrees with its specific context in two terms: physical-settlement, characterised by the relationship between architecture and space/scale/urban scene; relational context, represented by offerings and relations (functional and services) that welcome and accompany contemporary ways of dwelling and lifestyles (Oliva 2017).

We have various ways of imagining a project for the city of tomorrow. We can approach this as a nomad thinks about his steps as he crosses the desert: raising only the amount of sand necessary to move forward, leaving the minimum trace of our passage (conservation *tout court*); lifting the sand to observe where it is casually blown by the wind (aestheticising contemplation); lifting the sand to guide where it falls and build landscapes (informed and intentional design).

A good designer should probably adopt this latter approach when dealing with the system of open spaces. What is more, making operative use of the metaphor of "adaptative landscapes" (Wright 1932) to offer a diverse response to the surrounding environment, changes, unforeseen events and situations that may arise.

2.4 The Stations of the New M4 Metro Line as Neighbourhood Hubs

The study of the Green and Blue Backbone of the new Milanese M4 metro line reveals the possibility to transform the nodes, represented by its stations, into neighbourhood hubs, enriched by integrated and connected functions and public spaces that contribute to the creation of cities at the human scale. Hubs that also foster the rediscovery/promotion of the INA Casa/IACP[2] districts built during the 1960s and

---

[2] *Istituto Autonomo Case Popolari*, Autonomous Institute for Public Housing, an Italian entity created in 1903 to promote, build and manage public housing assigned primarily to low-income families, in particular under rent control schemes.

'70s, and support new forms of work based on the sharing of operational collective spaces located close to home.

The metro stations are signs/nodes in the territory that gather, distribute and bring flows to the surface. The stations of the new M4 line, part of a true and proper blue and green linear park, become urban thresholds (for the homes, services and activities they encounter). They are signs that characterise the present of the city. They are also enabling platforms for environmental (water, parks, landscaping) and infrastructural systems, but also its memory, intended as the history of the city, of its past, and a living memory, the culture the city produces, and its future, the transformations of tomorrow, which these stations must communicate and seek to guide.

How do these stations become thresholds of urbanity? By ensuring maximum continuity between pedestrian and bicycle connections between stations (city gates) that serve as attractors and generators of flows; by reducing and overcoming barriers that limit/impede maximum permeability/connectivity. Connections are obtained by overcoming/eliminating barriers, by creating new connections and/or improving existing ones, by enlarging pedestrian isochrones (to high and medium accessibility of 5 and 10 min), by bringing the stations closer to the districts they cross. Pedestrian and bicycle paths must branch out from the stations, innervating neighbourhoods and designing two different typologies of urban figures centred around the stations; 'stations-epicentres' in a 'constellation' of places; 'stations-trees' as the pattern of a network.

The project to contextualise the M4 stations began with a mapping of isochrones, of attractors and generators. Contextualising the stations of the new M4 line signifies first and foremost questioning the level of pedestrian and bicycle access to the stations themselves as a condition for verifying the catchment area of potential users, beginning with current conditions. The new stations are inserted in already dense urban fabrics, and thus with priority pedestrian access. Pedestrian accessibility was calculated by considering pedestrian isochrones of 5, 10 and 15 min (high, medium and low pedestrian accessibility). This allowed for a mapping of the catchment area of each station along the line.

The resulting map clearly shows the extent of the pedestrian catchment areas of each station, together with the obstacles (barriers, interrupted connections, etc.) that limit its fruition by pedestrians. The map offers suggestions about important intervention points through targeted actions focused on expanding each station's area of influence, securing pedestrian paths and resolving discontinuities along routes providing access to the stations. A condition of relevant interest to both the design and functioning of each station is represented by the distribution of uses that generate and attract flows of users and which fall within the area of influence of each station. This was followed by the mapping of attractors (functions that attract flows of users, including productive and tertiary-commercial activities and services) and generators (housing).

An analysis of the distribution of generators and attractors is a helpful indicator for estimating the potential flow of users of a station and for understanding temporal distribution during the course of a day. In other words, if the presence of generators activates demand concentrated prevalently during the peak hours of morning arrival and evening hours of leaving, specialised attractors present much more heterogeneous rhythms and, in general, are complementary to generators.

The mapping of generators and attractors also provides an important indication for orienting interventions to contextualise the stations as it offers indications about existing attractors that must be guaranteed accessibility to the M4 line, by securing and requalifying pedestrian paths providing access to the stations. It also permits an evaluation of flows potentially generated by and attracted to each station and the verification of conditions for eventual densifications by transferring new construction to spaces around the node. The mapping exercise considered the average value in Milan as the reference value for the three thresholds mapped (high, medium, low).

The following indicators were mapped for the attractors:

- 'population density' in each area of the survey, calculated as a ratio between population and area in this census district;
- density of pedestrian flows 'leaving' each section surveyed, calculated as the ratio between exiting flows and the active population of each census district.

The following indicators were mapped for the generators:

- 'density of workers' in each area of the survey, calculated as the ratio between workers and local units in each census district;
- density of pedestrian flows 'entering' each area of the survey, calculated as the ratio between incoming flows and workers in each census district (Figs. 2.1, 2.2 and 2.3).

References

Agnoletti C, Di Maio S (eds) (2011) Il contrasto alla rendita. Le nuove sfide dell'economia urbana. Supp 2, Scelte pubbliche, Associazione Romano Viviani, Florence

Alessandrini G (ed) (2019) Sostenibilità e capability approach. Franco Angeli, Milan

Anders G (1963) L'uomo è antiquato I. Considerazioni sull'anima nell'epoca della seconda rivoluzione industriale. Bollati Boringhieri, Turin

Angrilli M (2002) Reti verdi urbane, Palombo Editori, Rome

Arcidiacono A, Pogliani L (eds) (2011) Milano al futuro. Riforma o crisi del governo urbano. Et al/Edizioni, Milan

Arcidiacono A, Galuzzi P, Pogliani L, Rota G, Solero E, Vitillo P (2013) Il Piano Urbanistico di Milano (PGT 2012) – The Milan Town Plan (PGT 2012). Wolters Kluwer Italia, Milan

Arcidiacono A, Ronchi S, Salata S (2018) Un approccio ecosistemico al progetto delle infrastrutture verdi nella pianificazione urbanistica. Sperimentazioni in Lombardia. Urbanistica 159:102–113

Beck U (1992) Risk Society. Towards a New Modernity. London, Sage

Benevolo L (2011) La fine della città, Laterza, Bari

Bevilacqua G, Ricci L, Rossi F (2020) Rigenerazione urbana e riequilibrio territoriale. Per una politica integrata di programmazione e di produzione di servizi. In: Talia M (ed) La città contemporanea: un gigante dai piedi di argilla. Planum Publisher, Rome-Milan, pp 354–360

Blečić I, Cecchini A, Talu V (2013) The capability approach in urban quality of life and urban policies: towards a conceptual framework. In: Serelli S (ed) City project and public space. Springer, pp 269–288

Bonfiglioli S (1990) L'architettura del tempo, Liguori, Naples

Cacciari M (2004) La città, Pazzini, Villa Verucchio (RN)

Calderini M, Gerli F (2020) Innovazione, sfide sociali e protagonismo dell'imprenditoria ad impatto. Impresa Sociale 3:10-19

Camagni R (2012) Rendita immobiliare, tassazione e ricapitalizzazione delle nostre città. In: Cappellin R, Ferlaino F, Rizzi P (eds) La città nell'economia della conoscenza. Franco Angeli, Milan, pp 325–330

Camagni R, Modigliani D (2013) La rendita fondiaria/immobiliare a Roma: 6 studi di caso. Report presented at XXVIII Congresso INU, Città come motore dello sviluppo, Salerno, 24–26 Oct 2013

Campioli S (2020) Città inclusiva e senza limiti. Progettare luoghi per le persone nella società, Maggioli Editore, Santarcangelo di Romagna (RN)

Ciocca P (2021) Ricchi e poveri. Storia della disuguaglianza. Einaudi, Turin

Coppola A, Del Fabbro M, Lanzani A, Pessina G, Zanfi F (eds) (2021) Ricomporre i divari. Politiche e progetti territoriali contro le disuguaglianze e per la transizione ecologica. il Mulino, Bologna

Coupland A (1996) Reclaiming the city: mixed use development, Routledge, London

Curti F (2007) Lo scambio leale. Negoziazione urbanistica e offerta privata di spazi e servizi pubblici. Officina, Rome

De Carli et al (eds) (2011) PGT di Milano. Rifare, conservare o correggere?. Maggioli, Sant'Arcangelo di Romagna (RN)

Ferraro G (1998) Efficacia dei piani, efficacia delle teorie. Urbanistica 110, pp. 7–12

Gabellini P (2010) Fare urbanistica. Esperienze, comunicazione, memoria. Carocci, Rome

Gallino L (2011) Finanzcapitalismo. La civiltà del denaro in crisi. Einaudi, Turin

Galster G (2001) On the nature of neighbourhood. Urban Stud 38(12):2111–2124

Galuzzi P (2010) Il futuro non è più quello di una volta. La dimensione programmatica e operativa del progetto urbanistico. In: Bossi P, Moroni S, Poli M (eds) La città e il tempo: interpretazione e azione. Maggioli, Sant'Arcangelo di Romagna (RN)

Galuzzi P, Pareglio S, Vitillo P (2018) Città contemporanea e rigenerazione urbana. Temi, azioni, strumenti. Equilibri 1:125–133

Galuzzi P, Magnani M, Solero E, Vitillo P (2019) Residual Urban Spaces and new Communities of Social Practices/Spazi urbani residuali e nuove comunità di pratiche sociali. TRIA 12:31–50 https://doi.org/10.6092/2281

Galuzzi P, Solero E, Vitillo P (2020) Alpine Space Fragilities. A Research Line. Territorio 92:181–184

Galuzzi P, Vitillo P (2021) Trame resilienti per i territori della contemporaneità. Il caso della Città Vecchia di Taranto. In Brunetta G, Calderice O, Russo M, Sargolini M (eds) Resilienza nel governo del territorio, Atti della XXIII Conferenza nazionale SIU. DOWNSCALING, RIGHTSIZING. Contrazione demografica e riorganizzazione spaziale, Torino 17–18 June, vol 4: 10–16 Planum Publisher|Società Italiana degli Urbanisti, Rome–Milan

Gehl J (2011) Life between buildings: using public space, 1st ed. 1971. Island Press

Gehl J (2012) Vita in città, Maggioli Editore, Santarcangelo di Romagna (RN)

Giarelli G, Vicarelli G (eds) (2020) Libro Bianco. Il Servizio Sanitario Nazionale e la pandemia da Covid-19. Problemi e proposte, Franco Angeli, Milan

Karsten L (2003) Family gentrifiers: challenging the city as a place simultaneously to build a career and to rice children. Urban Stud 40(12):2573–2584

Krugman P (2008), La coscienza di un liberal, Laterza, Rome-Bari

Lavorato A, Galuzzi P, Vitillo P (2020) 8 Racconti di Milano. Ance, Milan

La Varra G (2007) Abitare in un loft nuovo, Multiplicity.lab. Milano. Cronache dell'abitare. Bruno Mondadori, Milan

Manzini E (2018) Politiche del quotidiano. Progetti di vita che cambiano il mondo, Edizioni di Comunità, Rome

Mareggi M (2011) Ritmi urbani, Maggioli Editore, Sant'Arcangelo di Romagna (RN)

Mariotti I, Di Vita S, Akhavan M (eds) (2021) New workplaces: location patterns, urban effects and development trajectories. A worldwide investigation. Springer International Publishing

Mattioli C, Renzoni C, Savoldi P (2020) La riapertura delle scuole. Una questione urbana'. La Rivista, Il Mulino, 2020 June 26. Retrieved from https://www.rivistailmulino.it/a/la-riapertura-delle-scuole-una-questione-urbana

Moreno C (2020) Vie urbaine et proximité à l'heure du Covid-19. Editions de l'Observatoire, Paris

Mumford L (2002) La città nella storia. Dalla corte alla città invisibile, vol 3, 1st ed. 1963, It. tran., Bonpiani, Milan

Nucci L (2012) Verde di prossimità e disegno urbano. Le open space strategies ed i local development frameworks dei 32+1 Boroughs di Londra, Gangemi Editore, Rome

Nussbaum MC (2011) Creating capabilities: the human development approach. Harvard University Press

Nussbaum MC, Sen A (eds) (1993) The quality of life. Oxford University Press, Oxford

Oliva F (2002) L'urbanistica di Milano. Quel che resta dei piani urbanistici nella crescita e nella trasformazione della città. Hoepli, Milan

Oliva F, Ricci L (2017) Promuovere la rigenerazione urbana e la riqualificazione del patrimonio edilizio esistente. In Antonini E, Tucci F (eds) Architettura, Città, Territorio verso la Green Economy. Edizioni Ambiente, Milan

Palermo PC (2011) Milano-Bigness. Quando la crescita non è sviluppo. In Arcidiacono A, Pogliani L (eds) Milano al futuro. Riforma o crisi del governo urbano. Et al./Edizioni, Milan

Pareglio S, Vitillo P (2013) Milano. Metabolismo urbano nella città ordinaria. Urbanistica 152:65–73

Pasqui G (2008) Città, popolazioni, politiche. Jaka Book, Milan

Pievani T, Varotto M (2021), Viaggio nell'Italia dell'Antropocene. La geografia visionaria del nostro futuro. Aboca Edizioni, Sansepolcro (AR)

Rilke RM (1980) Lettere a un giovane poeta. Adelphi, Milan

Ross A (2021) I furiosi anni Venti. La guerra fra Stati, aziende e persone per un nuovo contratto sociale. Feltrinelli, Milan

Saint-Exupéry A de (1951) Cittadelle. Paris, Gallimard

Sapelli G (2011) Tra rendita e rendita finanziaria: la città a frattali. Scelte pubbliche 2:9–16, Associazione Romano Viviani, Flornce

Saraceno C (2021) Il welfare. Tra vecchie e nuove disuguaglianze. Il Mulino, Bologna

Schumacher EF (1973) Small is beautiful. A Study of Economics as if People Mattered. Blond & Briggs Publisher, London

Sennett R (2012) Insieme. Rituali, piaceri, politiche della collaborazione. Feltrinelli, Milan

Solero E, Vitillo P (2021) Territori fragili al centro. Le aree interne, luoghi da riabitare. Territorio 97:132–137, Special Issue

Stiglitz J (2002) Crony Capitalism American-style. Project Syndicate, 11 Feb

Tajani C (2021) Città prossime. Dal quartiere al mondo: Milano e le metropoli globali. Guerini e Associati, Milan

Tamini L (2020) Distretti del commercio e nuova pianificazione urbanistica. In: Franco E, Tamini L, Zanderighi L (eds) Commercio e Distretti: un patto per lo sviluppo, Maggioli Editore, Sant'Arcangelo di Romagna (RN), pp 23–30

Thaler RH, Sunstein CR (2008) Nudge: Improving Decisions About Health, Wealth, and Happiness. Yale University Press, New Haven, Connecticut

Tonon G (2014) La città necessaria, Mimesis, Sesto San Giovanni (MI)

Vitillo P (2011) Gli uomini, non le case, fanno la città. Lo Squaderno 21:25–32

Vitillo P (2022) Nuove forme di urbanità per la città contemporanea. Urbanistica 164:4–7

Zajczyk F (2007) Tempi di vita e orari della città. La ricerca sociale e il governo urbano. Franco Angeli, Milan

Zucchi C (ed) (2014) Innesti Grafting. La Biennale di Venezia. 14. Mostra Internazionale di Architettura, Marsilio, Padua-Venice

Zucchi C (2021) Una città (non) è un albero. In lode di un disegno urbano just-out-of-time. Equilibri 2:421–430

Wright S (1932) The Role of Mutation. Imbreeding, Crossbreeding and Selection in Evolution, Proc. 6th Intern. Cong. Genetics 1:356–366

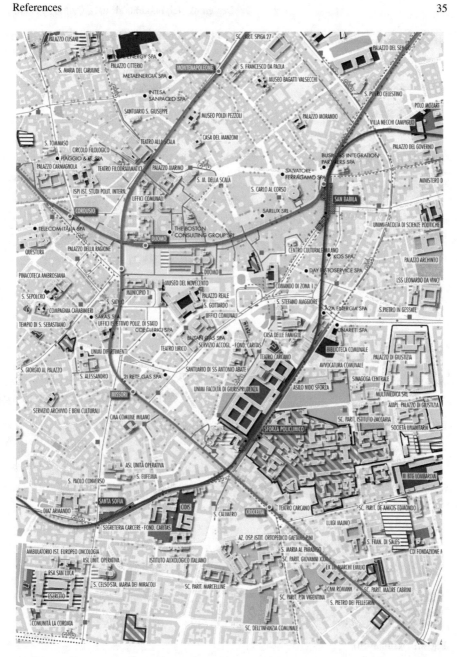

Fig. 2.1 Analyses of the urban context of the Sforza-Policlinico station. Original drawing at 1:5000

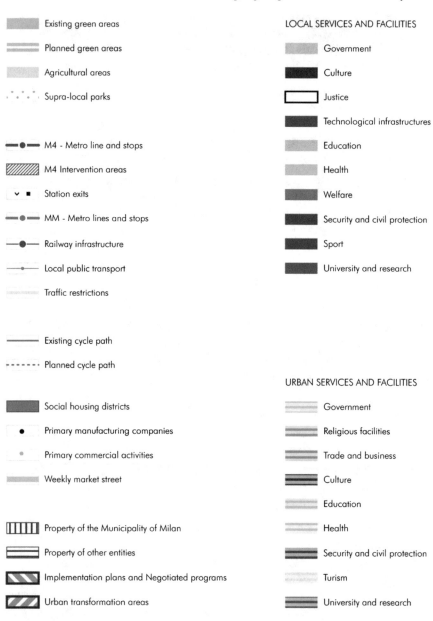

Existing green areas

Planned green areas

Agricultural areas

Supra-local parks

M4 - Metro line and stops

M4 Intervention areas

Station exits

MM - Metro lines and stops

Railway infrastructure

Local public transport

Traffic restrictions

Existing cycle path

Planned cycle path

Social housing districts

Primary manufacturing companies

Primary commercial activities

Weekly market street

Property of the Municipality of Milan

Property of other entities

Implementation plans and Negotiated programs

Urban transformation areas

LOCAL SERVICES AND FACILITIES

Government

Culture

Justice

Technological infrastructures

Education

Health

Welfare

Security and civil protection

Sport

University and research

URBAN SERVICES AND FACILITIES

Government

Religious facilities

Trade and business

Culture

Education

Health

Security and civil protection

Turism

University and research

Fig. 2.1 (continued)

Fig. 2.2 Interpretative summary of the urban context of the Sforza-Policlinico station. Original drawing at 1:5000

Fig. 2.2 (continued)

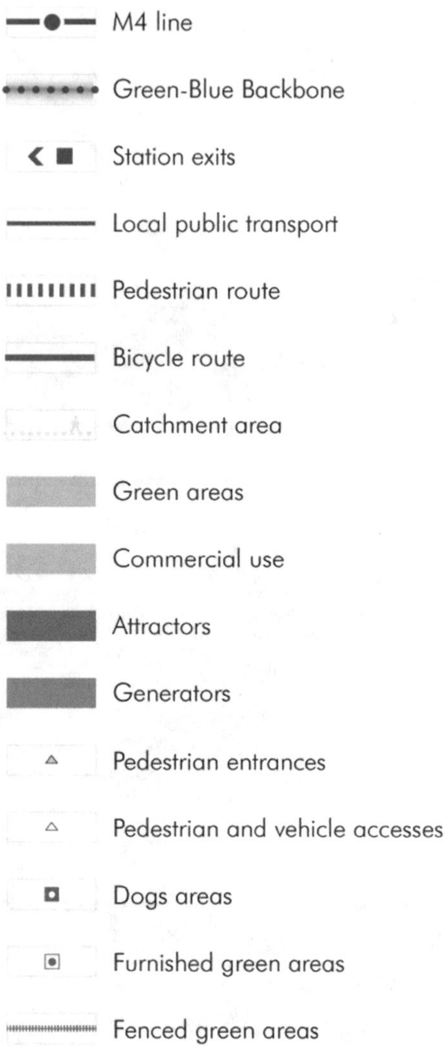

—●— M4 line

●●●●●●● Green-Blue Backbone

◀ ■ Station exits

—————— Local public transport

IIIIIIIII Pedestrian route

—————— Bicycle route

⋯⋯⋏⋯ Catchment area

Green areas

Commercial use

Attractors

Generators

△ Pedestrian entrances

△ Pedestrian and vehicle accesses

▣ Dogs areas

◉ Furnished green areas

▥▥▥▥▥ Fenced green areas

Fig. 2.3 Attractors (in orange) and generators (in lilac) of users of the M4 line and potential users of the green–blue backbone

Chapter 3
The Re-signification of the City and Inherited Building Stock

Abstract The re-signification of historical heritage is one possible action for triggering real processes of urban regeneration in the contemporary city. The historic city represents the concentration of the know-how and values that have come to define and characterise the evolution of society and the environment it inhabits. During the sixty years since the approval of the first Gubbio Charter (1960) for the conservation of historic centres, urban historical heritage has often represented an impediment and a limit on the natural evolution of the city. Many of the current reflections on urban regeneration consider the historic city as something to be defended and protected against the impoverishment of the landscape and the territory. While this may be undeniable, in this chapter, we will underline how historical heritage represents the resource from which to set out and define a process of pragmatic regeneration, though based on the recognition (or definition) of new local identities (of proximity). A similar approach overcomes the dichotomies between conservation and transformation and includes proactive attitudes toward understanding and caring for the territory.

In this chapter, proximity is defined as the 'expression of identity' and the paragraphs that follow reflect on three themes. Firstly, the theme of urban history concerning the dimension of the contemporary city, as a revealing of the values of a community, redesigned beginning with the new geographies of ancient and modern heritage at the territorial scale. The historical territory is intended here as a structural network of the city that consolidates bonds of proximity, memory and identity of communities. The second theme centres on the street as the arena of a sedimentation of relations between space and society, between heritage and community, between conservation/protection of identity and the valorisation/promotion of urban history. The third theme regards tourism as a tool for rediscovering historical cities to keep them from succumbing to either 'hit-and-run' tourism or the process that re-directs the vacation market toward destinations closer at hand, less familiar and less crowded. The chapter concludes with a return to a number of aspects of the design of Milan's Green and Blue Backbone that, by thematising the public spaces around the new stations of the M4 metro line, restores the *urbs* to the *civitas*.

© The Author(s), under exclusive license to Springer Nature Switzerland AG 2022   41
M. Fior et al., *(Re)Discovering Proximity*, PoliMI SpringerBriefs,
https://doi.org/10.1007/978-3-031-08958-9_3

3.1 Proximity as Identity. From the Historic Centre to the Historical Territory

3.1.1 Calling Things Back to Memory

In the social sciences, the concept of identity explores how people perceive themselves and their place in society and what characterises and sets them apart from others. Identity is not immutable but transforms as society develops and changes. Able to observe the world, but not one's self, an individual must make recourse to tools that allow for self-revelation. As a physical object, the city is one of these instruments of self-discovery and self-comprehension, above all when the subject is not a singular individual but a community.

Simmel (1908) assumed space as a condition for the existence of communities, in other words, he considered the urban environment not as a backdrop against with society acts, but instead as a property of society. Hence, through the observation of the city it is possible to define a collective identity by recognising the signs that link a community to a place, in other words, its *genius loci* or "statute of place" (*statuto dei luoghi*), as indicated by the Territorialist approach (Magnaghi 2000). Collective identity is the conception a community has of itself with respect to its founding values, concretely represented in the rich 'palimpsest' of the territory (Corboz 1985). It follows that the city conserves a 'historical territory'—the result of a constant process of stratification and writing by natural processes and human interventions—to be revealed through analysis and design at the real scale of the metropolized contemporary city.

The revealing of the historical territory, and its use as a tool for the design of the city, is an attitude that consolidates collective memory and identity by attributing new meaning to the sum of historical traces. The concept of the 'historical territory'—born in the early 1990s in Italy (ANCSA 1990)—testifies to a new urbanist culture that makes every effort to work in the present, even with the faintest and most unusual traces of the 'only recently cooled past' (Gasparrini 2008). The idea is to delineate the future traits of the city and society within a continuous process of re-signification that no longer sets any limit, boundary, or separation between materials of value and the de-qualified city.

There exists today a new awareness of the role of urban history and its typo-morphological expressions, derived from various processes, such as, for example: the sensibility of societies toward themes of the environment and health and safety deriving from the unsustainability of the metropolized city (Oliva 2010); the presence of new ties and a sense of belonging to the territory that have emerged from the study of current practices of using the city (Cellamare 2009); the abandonment of many stereotypes, such as, for example, the ugliness of the short twentieth century and its architectural, urban, industrial and artistic expressions (Di Biagi 1999). Indeed, in virtue of the territorial, social, and economic changes that have come about, which can place material and immaterial heritage at risk, that we see a powerful re-emergence of the need to develop new awarenesses, policies, and actions to safeguard the historical city.

The contemporary historical framework is composed of an articulated repertory of heritage recognisable at various scales: from medieval fortifications to Renaissance palaces, from archaeological remains to Rationalist neighbourhoods, from late medieval agrarian land divisions to Baroque gardens, from the roads built by the Romans to the railways of the late nineteenth century. This variegated system of open and built spaces is characterised by intrinsic and relational values revealed by urban design. They have documentary and historiographic value and blossom to produce other values: environmental, ecosystemic and landscape-related, as well as ludic-recreational and mnemonic.

The act of remembering, of calling things back to memory, is fundamental to identity construction, and partially determined by the vision and knowledge of existing signs and traces. This does not imply that cancelling historical traces always negatively affects the affirmation of a specific identity. Instead, it signifies that the conservation or suppression of the territorial palimpsest comports a re-elaboration and actualisation of its significance. When a community re-signifies traces, it determines a mechanism of appropriation of places that defines a new relation between space and society. For this reason the historical city is an 'identifying structure' of place (Manieri Elia 2008). In other words, it can demonstrate the distinctive traits of a community by placing it at the centre of a project for urban regeneration.

The current urban condition—characterised by flows and uses that have caused an implosion in proximity spaces—has dematerialised the anthropological premise of the *civitas*, that is, the capacity to edify the *urbs*, intended as its living environment. However, thanks to an intelligent work with the materials of history (*pars construens*) society can once again appropriate the skills of the art of building (*l'art d'édifier*) its living environment (Choay 2008). An environment at the human scale, characterised by the proximity of diversified functions and rediscovered or new values, as well as a stimulus to the "cooperation as a factor in the evolution of society itself" (Nowak 2006). Attentively selecting the elements to be preserved legitimises a continuous remodelling and stratification of places, avoiding museum-like conservation, the mercification of heritage and the consequent alienation of the communities that inhabit them. Selecting aids memory, and memory aids the transmission of "urban facts" to future generations.

The "urban facts" referred to by Rossi (1995)—streets, quarters, buildings—can be considered works of art because "all great manifestations of social life and all great works of art are born in unconscious life". Cities reveal the temporal experiences that have societies have practiced unconsciously within and with them. For this reason, cities (or at least parts of them), can be considered 'works of art' *tout court*; in the sense that they are influenced by the territory in which they are born, but at the same time condition the evolution of this context and the society that inhabits them.

Not by chance, every urban element inherently contains a memory recounted through a form. Urban forms, revealed like works of art, constitute a value more powerful than the environment in which they are located and more resistant than memory. It is because the value they conserve is the idea of the city that generated them; they are the exemplification of how the structure/substance of a good is bound up within its form. Remembering is an action that takes place in the present, making it strictly dependent on the context in which this action occurs. The reconstruction of the past corresponds with present-day society's interests, ways of thinking and ideal needs. Remembering thus signifies actualising the memory of a group. The image that derives from the past and that memory actualises is not something immutable, nor something definitive: the past is conserved in the life of men, in the spaces they have experienced, and the forms of consciousness these spaces have generated them. "Collective memory" consists of a mass of dense and mobile memories that are continually modifiable and re-buildable based on the necessities of living and active social groups (Halbwachs 1987: 28).

When returning to a city visited in the past, what we perceive visually aids the reconstruction of a set of forgotten events. On the one hand, what we observe is repositioned in old memories; on the other hand, memories adapt to changes to what we maintain in the present. The concordance of this mnemonic feedback makes it possible to reconstruct a set of recognisable memories (a collective memory). If we find that a place once visited has changed considerably after many years, we fail to recognise it, but we also modify memories that are completely annulled. What we see as totally new has no place in the memory of what was, while we adapt fully to changes in what we see in the present. This means that the complete annulment of the space in which memory re-emerges also annuls the memory itself. Though not necessarily all elements, as the permanence of a few is sufficient for memory to be regenerated.

Memory is configured as a weak human attitude if taken singularly. That is, not only remembering and memory can be annulled in the absence of direct evidence, but there is also a risk of annulment when memories are not shared. Hence the importance of the "collective" dimension. "Collective memory" makes it possible to transmit a group experience through time. Collective memory is structured when multiple individuals belong to a group and, in a certain sense, think together. Remaining in contact with one another, they can identify themselves with this group, completely fusing the past of a single individual with the group's history. Memory is formed through collectivity because the single memory disappears and has belonged to the group that conserves and transmits it for some time. For individual memory to survive it is not enough that the group presents its own evidence; individual memory should match collective memory and there must be points of encounter whose re-evocation lays the foundations for this common structure.

Collective memory is profoundly social in nature because it links the sharing of an experience. By expanding this reasoning to the city we can consider it the most essential space of human experience, the place where shared collective action occurs, if not for direct reasons, at least for indirect one linked to the proximity of common spaces. Given the collective nature of memory, preserving (public) urban spaces is not a process that tends to generalise all urban palimpsests. Instead, it selects, in virtue of a broad sharing of recognition, only those spaces thanks to which society manages to recognise itself, and decides to conserve for future generations. In this sense both voids and solids, presences and missing elements, are eloquent. This is a means for organising and structuring more or less shared repertoires of spaces, formed and modified over time, becoming places, discarding certain presences and privileging others. This mechanism which alternates remembering and forgetting generates identity (Helzel 2016).

The relationship between the form of the city and the society that inhabits it is reinforced by the mnemonic bond through which collective society associates experience with place. Thus, there is no collective memory that does not unfold also in a spatial framework. Space is a shifting reality, but it persists: we could not find the past were it not physically preserved, even in the minimum doses intent on continuing it, in the material world that surrounds it. The space occupied by humankind must be the object of the attention of urban design because it is the direct custodian of collective memory.

3.1.2 The Italian School: From the 'Historic Centre' to the 'Historic City'

By involving heritage, the process of historicity permits an evolution in continuity with the past. In other words, the historicity of a piece of heritage is its character of 'becoming historical' over time, admitting mutations and evolutions. Opposite to it, Historicism is the interpretation and evaluation of this same element concerning the present moment and the historical environment in which it was produced.[1] Dealing with historicity in disciplines concerned with space signifies working with design: for a monument, for a group of buildings, for homogenous/recognisable portions of the city, until we comprehend the existing city in its entirety (Gabrielli 1993; Bonfantini 2002). The history of a place and its community is manifest and described through "urban facts" (Rossi 1995) that in turn become the object of different interpretations, descriptions, and attitudes for their preservation. The techniques of urban planning employed to investigate historicity in the city have sedimented approaches to preservation that—beginning with interventions of conservation reserved for the most ancient parts of the city—were later implemented through adaptations and contaminations, and applied to all that was "consolidated within a territory" through a "discrete regulation" (Bonfantini 2002). The evolution of approaches to conservation has revealed the expansion of the set of possible interventions for regulating historical fabrics (from isolation to rehabilitation, from conservation to requalification). On the other hand, it has raised the necessity of preserving urban historicity (from the historical monument to urban environments, from the historic centre to the historical territory), admitting an expansion of borders to identify heritage of historical interest.

When we refer to the historical city, everyone has their idea or point of view on what it is. This construct tends to be associated with a rather circumscribed image:

---

[1] Definitive from the *Dizionario Hoepli Italiano* by A. Gabrielli, available via *eLexico* from the website of the Politecnico di Milano.

a warren of narrow streets, a few widenings, or small public squares lined by houses and shops that define a single organism, distinct and homogeneous, and referred to in urban planning terms as the 'historic centre.' Historic centre because on the one hand, this structure of settlement is generally located in the heart of the city (at its centre) and, secondly, because the spatial forms described by this term belong to a way of conceiving of and building the city that has now been surpassed (historical because it refers to the past). In reality there is a notable cultural distance between what is described by the notion of a 'historic centre' and that of a 'historic city.' A depth of thinking and an evolution of techniques of urban-building intervention with relatively distant origins and a process of slow and often contrasted formation. An evolution that has traversed centuries and which today appears to have found a new codification in the concept of the 'historical territory' (ANCSA 1990). The reflections and experiences of the past sixty years in the fields of recovery and conservation of historical urban heritage have now consolidated the idea that the historic centre, circumscribed within the physical perimeter of the boundary of an ancient city it is not sufficient to gather up the dense and layered, living and active memory of a place and a community. Despite this, the historic centre plays an influential and symbolic role in identifying the privileged space in which the most relevant historical, artistic and cultural values to be preserved are concentrated. Today there is a consolidated idea that 'historical quality' must be recognised in cities or territories whose dimensions exceed those of an ancient city.

The cultural passage from the concept of the 'historic centre' to that of the 'historical city' contributes to the abandonment of the functionalist approach to zoning (DI 1444/1968 instituted the *A Zones* that include "those parts of the territory home to urban agglomerations of a historic or artistic character or of particular environmental value, or portions thereof, including surrounding areas, which can be considered an integral part, owing to these characteristics, of these same agglomerations"), or which identify the historical memory of the entire body of the existing city (through the institution of 'urban fabrics'). Admitting the evolution of urban historicity, this passage exalts both the values of ancient nuclei, as well as the values of modern (and even contemporary) architecture and urbanism, that is, places with a recognised symbolic, testimonial and cultural value, and thus of identity for the city and society. Values that are not read through differences and oppositions but in continuities, successions, and evolutions. This important methodological leap forward overcomes the historical dilemma between conservation and transformation, two tendentially opposing attitudes, and give rise to conceptual and operative difficulties (Cervellati 2010). Nevertheless, in Italy, when we speak of 'urban fabrics' we refer primarily to those parts of the city that are homogenous by typologies, morphologies, building proportions and spatial relations, history, era of construction and successive stratifications, including functions and use's that, while diverse, are also compatible with one another. Among urban fabrics, we tend to single out those with an ancient history and those with a more modern one. The first refers to the area of the 'historic centre' for which there continues to be a necessity and urgency to enact unquestioned preservation (Guermandi and D'Angelo 2019). Modern-contemporary fabrics, instead, belong to that part of the existing city with a recognised historical value for its being the result of a process of urban planning. Through their recognisable urban design, these urban fabrics of the 'historical city' reveal the vision of the development of settlement typical of the period spanning between the late 1800s and the post-war era. In other words, these parts of the city are configured as the legacy of the plans created by modern urbanism.

The concept of the 'historic centre' is the result of an evolution of the discipline that recognised the historical and cultural value and identity of an organic part of the city, and not only monuments and their immediate surroundings (Fior 2020a; b). With the passage from the conservation of monuments to the conservation of the historic centre we can mention the numerous essays by Françoise Choay (1973, 1992, 1995), though in particular the debate among Italian urbanists in the wake of the definition of the Gubbio Charters, by the Associazione Nazionale Centri Storico-Artistici. The first Gubbio Charter (1960) sanctioned the existence of a "single monument" to be preserved: the historic centre, made of exceptional works of architecture, as well as minor works of architecture and the public spaces linking them. These buildings and spaces are substantially circumscribed within the expansions of medieval city walls (in smaller towns) or seventeenth/eighteenth-century walls (in larger towns). The second Gubbio Charter (1990) introduced the concept of the 'historical territory,' which expanded the paradigm of the historic centre. Article 2 of the second Gubbio Charter reads: "In every European city the historic centre has represented the area in which the values of the *civitas* and the *urbs* have been concentrated: its protection and promotion are necessary to guarantee the historical identity of settlements and thus their value. The historic centre also constitutes the node of a vaster structure of settlement. This structure, interpreted in its centuries-long process of formation, must now be identified as a 'historical territory,' the comprehensive expression of cultural identity and thus subject in all of its parts (existing city, built landscapes, rural territory) to an organic strategy of intervention".

In reality, in Italy, the second Gubbio Charter from 1990 anticipated UNESCO's recommendations for the Historic Urban Landscape (HUL 2011). It is defined as follows in art. 8: "The Historic Urban Landscape is the urban area understood as the result of a historic layering of cultural and natural values and attributes, extending beyond the notion of 'historic centre' or '*ensemble*' to include the broader urban context and its geographical setting".

In a 2012 interview, Bruno Gabrielli stated: "The UNESCO Declaration on the theme of the Historic Urban Landscape contains two innovations. The first is the recognition of 'immaterial, and not only 'material' heritage, among that to be preserved. Or the comingling of material heritage and immaterial value. For example, a poem could bring value to a wall, a building, an urban perspective... and this is very important for defining that which possesses a historic and cultural value and identity. The second innovation is that the UNESCO recommendations place a strong accent on the contemporary. Heritage conservation is guaranteed when it is contemporary, that is, if the value of this heritage is recognised in our contemporary era. We consider heritage actual if there have been or exist hypothesis for its re-contextualisation. And design offers a means for making inherited heritage contemporary" (Fior 2013: 120).

The second Gubbio Charter aimed at being a written contribution to building consensus and sharing decision-making processes for confronting the regulation of the historical and cultural heritage and identity in the city and territory in a more structured manner. These rules were to uniform both the technical language employed when approaching the notions of 'conservation' and 'transformation.' Still, above all they were to have unified the methods used to investigate and recognise the founding characteristics of the existing city and built landscape. This sanctioned the importance of a "project of knowledge", that is, the process of understanding the potentialities of settlement that are the foundation of any urban regeneration project (Mazzoleni 1991: 36).

As Bruno Gabrielli wrote, the 1990 Charter "is an entirely open document, even incomplete, but which contains the sum of the positions of ANCSA[2] matured during the first 30 years of its existence. Here we reflect firstly on the failure of the hypothesis of public intervention [in the field of preservation], the affirmation of the principles of protection but also the need for innovation and, above all, there is a consolidation of the idea of a strategy that regards not only the historic centre, but also the exiting city" (Gabrielli in Toppetti 2011: 12).

The 1990 Gubbio Charter definitively sanctioned the presence of the historical value of cultural and natural heritage located outside the perimeter of the historic centre. At the same time, the second Gubbio Charter also sanctioned the dilatation of the meaning and the field of intervention of preservation: from an action of conservation increasingly less tied to the defence of existing values toward an innovative action, increasingly more founded on design as the "privileged space for the production of the new values of contemporary society". Indeed, this outlined "a new philosophy of behaviour toward historic and natural heritage and its relations with the territories of our contemporary era, destined to have an impact on conceptions of the city, historic centres and the landscape, overturning many consolidated divisions and opening up expectations for reform" (Gambino in Volpiano 2011: 19).

The passage from 'historic centre' to 'historic city' or 'historical territory' has introduced important innovations in the conservation of urban historicity. First and foremost, historicity becomes a theme of urban planning, which determines an expansion in the scale of intervention of preservation (from the single historical monument to a more complex system of urban materials). In particular, the conservation of urban historicity occurs by maintaining and consolidating relations between different parts: between different historical values (ancient, modern, and even contemporary eras) and between "urban facts" and the community. The expansion of the list of "urban facts" with a recognised historical value, transforms the 'conservation of heritage' into the 'conservation of the relations between heritages,' which translates into the construction/design of 'networks di historical values.' We are dealing with a system of relations that does not negate, but instead complements and increases the historical

---

[2] The Italian National Association of Historic-Artistic Centers (*Associazione Nazionale Centri Storico-Artistici*, ANCSA). The ANCSA was founded in 1961 to promote cultural and practical initiatives to support the activities of public administrations in safeguarding and regenerating the urban heritage.

interest in traditional heritage. The awareness that historicity traverses the entirety of the territory, emerging in signs and traces of differing typology, location, date, etc. What is more, it comports the application of widespread, often ordinary, interventions, no longer focused exclusively on a special/exceptional spatial environment. Finally, the action of preserving historicity implies recognising the mnemonic value and identity of historical-artistic urban materials, attributing a social meaning to conservation. The idea of conserving urban historicity comports an admission of the evolution of the facts of which it is comprised. In other words, the transformation (and even the elimination) of some traces to respond to the necessities and meanings attributed by contemporary society to those parts held to possess value and that it wishes to pass on to future generations.

For these reasons, the concepts of 'historic city' and 'historical territory' can be considered largely equivalent. Underlying both is the premise of recognising the complexity of traces, in the various forms and eras they represent, and the ties they establish with society. The possible distinction between the two terms is that while 'historic city' refers to a collection of signs that belong prevalently to built agglomerations; the term 'historical territory' refers also to the traces of open spaces (nature and landscape) and the territorial morphologies that structured the environments in which communities live.

3.2 A Framework for Regeneration. Networks for Structuring Neighbourhood Identities

3.2.1 The Historical Territory Network

The Covid-19 pandemic accentuated the challenges for the cities of the future, adding health issues to the sweeping changes to be governed, which already included the environmental, climatic, economic, social, digital, and technological changes taking place. At the same time, we are witnessing a process of redefining urban planning (its tools, approaches, mechanisms and issues) that receives questions and demands for solutions to adapt cities to global changes. A European Union document from 2011 shed light on the fact that growing populations in the world's metropolises are accompanied by phenomena of social segregation in urban areas and the depopulation of many peripheral territories and historic landscapes. This second aspect, in particular, has exposed the many intrinsic vulnerabilities of territories, fragilities linked to the natural risks inherent to specific sites (earthquakes, landslides, flooding, etc.), outdated infrastructures (roads, railways, and information networks), and the absence of accessible, structured and quality social welfare.

Despite this, cities play a fundamental role as economic drivers, sites of connectivity, creativity and innovation, and centres of service. Since the 1990s, however, their administrative boundaries have ceased to correspond with the physical, social, economic, cultural or environmental reality of traditional development. There is a need for new ideas to concretely implement sustainable urban regeneration projects. By promoting innovation, we can support the transition (not only ecological) of the cities of the future, following the principles of the Urban Agenda EU 2030, reiterated in the Leipzig Charter (EC 2020), which promotes more ecological, inclusive and cohesive, productive and connected cities. For this reason, digital, social, environmental, climatic, health-related or economic challenges must be confronted at diverse scales and in an integrated manner. We must pay attention to the needs of quarters (to increase energy performance, support the protection of historical values-identity, and reduce phenomena of social segregation) and those of metropolitan territories. Guaranteeing the coherence between sector-specific initiatives focused on providing more efficient accessibility to services, mass mobility via public transport, the recognisability of places and biodiversity preservation.

Nonetheless, before producing any vision of the future, there is a need for solid know-how to support a shared comprehension of the potential to regenerate cities. A fundamental role in this perspective is played by the elements that structure and innervate new immaterial relations and physical connections in the contemporary city. These networks are the only components that support a process of spatial reconfiguration that goes beyond the infinite extension of the contemporary city (Bonomi and Abruzzese 2004). Networks overcome the discontinuities, the porosities and fragmentations (natural and socio-economic) created in the city and surrounding contexts. Not least, historical heritage, as a system of historic-cultural values and identities, characteristic of contemporary territories, is added to consolidated social, infrastructural and transport, environmental and ecological networks (Secchi and Viganò 1998; Campos Venuti 2001; Oliva 2001).

The objective is to work toward the common good and public realm, focusing on citizens, entrepreneurs, institutions and the new roles of administrations in governance. As well as overcoming the most urgent challenges, such as social housing and inclusivity, attractiveness for businesses, preservation of historical-cultural heritage and identity and ecological-environmental sustainability. In this vision of the future of cities, public space is no longer simply the space between buildings, but a space that generates a new urbanity. Urbanity is no longer bound to the codified and reassuring area of the city centre, or an elevated density of buildings, but covers a widespread condition of contemporary dwelling, with different lifestyles and expressions of historicity.

The regeneration of the contemporary city is substantially different, despite representing a normal evolution of the era of urban requalification that characterised the strategic programming of Italian cities since the 1990s. Urban regeneration obliges us to confront new lifestyles, new necessities (social dwelling) and the scarcity of resources (economic, but also the exhaustion of environmental resources, such as the soil or energy). The first goal of regeneration is the restoration of an equilibrium across the entire territorial system. For this reason, any new strategy of territorial governance must be developed under the banner of urban regeneration. Urban planning—only one of the many components of territorial governance—thanks to its specific tools (the master plan and the urban project), can provide the framework of territorial coherence within which to develop each new hypothesis for the regeneration of the environments in which we live. Within this general framework, preparing a project for the city that considers the 'historical territory network' helps reinforce a vision of urban development founded on elements that identify a place and configure a regenerative process rooted in the community. In other words, the historical territory is one of those networks that bring structure to the city of the future, beginning with its relations with the communities that inhabit it.

The concept of 'historical territory' is not intended to substitute that of 'landscape.' While both concepts lack a precise physical or temporal dimension, replicable in any context, the landscape is a 'non-fractionable unit' because it is characterised by an indissoluble dimension of spatial–temporal evolution. On the contrary, the historical territory is a 'selection of units,' of territorial elements that make up its structure, soul, genesis and identity. The historical territory is similar to a selective network of nodes connected by material and immaterial relations. Because of its selectivity, the historical territory doesn't exist as an open entity. Instead, it reveals itself and its components only through a 'project of understanding' because the historical territory is first and foremost symptomatic of the elements of which it is composed. It represents the moment when historical components are investigated. Secondly, the design of the historical territory network explicates the relations a community establishes with and through the nodes of this network. In other words, despite both terms being dynamic interpretations of space and its mutations over time. The landscape is nonetheless 'a system of ecosystems' (the adhesive, the amalgam), while the historical territory is 'a system of persistences,' that is, a network of elements that persist over time and to which society attributes meaning, or is willing to do so.

Therefore, the elements that belong to the network of historical territories are not only those that persist, as morphologically identifiable signs in the territory, but are also those that persist over time. They acquire a role in structuring the collective identity, even when it is modified and adapted.

The key point of the passage from preserving historical heritage to a regenerative project for the contemporary city—through the re-signification and creation of a network of the signs of urban history (above all in those contexts that, for various reasons, have ignored or even lost the distinctive signs of a common history)—occurs only when two operations develop in parallel. On the one hand, the formation of a 'collective conscience' that collaborates from the bottom-up in the search for the distinctive signs in territories, actualising their meaning and significance. The subjective value that each person may attribute to an element of heritage is born of countless motivations bound to personal experience: from the emotions for a birthplace or space in which we spend a great deal of time, but also through literature, music, poetry, with nurture the emotions of individuals toward an element of heritage. This emotivity can involve many subjects, and in the end, the faithful preservation of historicity, a strong guarantee for the preservation of heritage, lies in the fact that this element is recognized, independent from the motivations of a large part of society. On the other hand, there is the need for the formation of a 'technical conscience' that, from the top-down, through specialised know-how, gathers the signs and traces of historic interest and repositions them within a framework of urban coherence, capable of generating a new model of settlement, unified and functional, for the city. Substantially at least two conditions must be verified if a project for the historical territory is to acquire concreteness and have an effect on regenerative processes: citizens must appropriate the meaning of particular elements of historical interest that make the population an actor in the territory, capable of caring for the traces and values it preserves; and, on the other hand, there must be a technical know-how (typical of urbanism) that establishes a network from the many signs that express urban historicity and at the same time are capable of synergically attracting resources, actors, energies and ideas for the regeneration of the contemporary city.

The historical territory is a complex system of elements that delimit any project for the regeneration of the contemporary city. By interacting with and reinforcing itself through connections with other structural networks (environment and mobility), the historical territory can bring new quality to metropolized, fragmented and discontinuous urban systems (Fior 2013). The historical territory is a sum of goods that often coincides with the public city (publicly owned and used properties and services). The 'public city'[3] is historically deputised with collecting and layering the life of communities in its public squares, civic palaces, places of worship and streets. In this overlapping of functions and values, the historical territory becomes an 'infrastructure' for generating new urbanity (Gabellini 2010; Bonfantini 2013).

3.2.2 Networking (Historical) Public Spaces

Urban regeneration presents challenges that are political and economic, technological and linked to social innovation. It requires the development of new supply chains, experimentation with new approaches and the activation of new business models, not to mention an ever closer interaction between public and private subjects. In this situation, the concept of 'resilience' constitutes "a fundamental reference for the development of an idea and practical application of urban planning oriented toward the future" (Talia 2017) and the field of experimentation for planning in the implementation of sustainable urban regeneration (Musco 2016). As part of changes to society, the concept of resilience supports the affirmation of new inclusive models of coexistence, the restitching of relations with the territory, based on the management of risk and the promotion of the landscape, and the adaptation to global changes through new, lasting circular green economies.

Above all, regeneration is a theme of urban design, with evident effects on physical space, which must be guided (that is, planned) to be tangible and efficacious. The approach supported here is one of practicing urban regeneration through the identification, design and management of a "frame of networks" (*rete di reti*) (Ricci et al.

---

[3] In Italy, the term 'public city' refers to all the areas, spaces, services and public or community facilities that make up the city. It is a highly varied territorial endowment that includes schools, health facilities, parks and gardens, public buildings and places of worship; but also streets, squares and car parks. Town planning generally considers the areas for urban standards regulated by Decree n. 1444/68, together with a more articulated welfare system (e.g., social housing districts).

2018). In particular, blue-green networks and historical heritage become structural and structuring elements of possible strategies for the regeneration of the contemporary city. In these networks there is a linking, an overlapping, intersection and operativity of open and relational spaces functional to the inhabitability and recognisability of places.

The concept of the network has dilated and assumed diverse meanings. In scientific language, the term is employed in diverse fields, from economics to computer sciences to ecology. In sociology, for many years now, networks have been used as important elements in processes that aid and promote the quality of life. In particular, in social studies, fundamental networks include those that unite people through blood ties, friendship, neighbourhood relations, and which naturally and reciprocally support one another. However, with the term network, we also refer to the constitution of relations among institutions (governments, agencies, public and semi-public entities) or the collection of people who gather, informally, to seek solutions to common questions (associations, NGOs, committees). In urban studies, the theme of the network is widely used above all in the field of transportation and when discussing the offering of underground utilities, for example, the supply of electrical energy or water, where we are dealing with sectors that supply the territory with material services. Despite this, the theme also developed toward immaterial services, that is, by leveraging the system of values (information, resources, skills) that can be connected between cities or within them, positioning urban agglomerations, or parts thereof, within a hierarchy of roles and nodes, and setting them before the choice to collaborate with one another to ensure their survival in the era of globalisation.

The conditions of the contemporary city (boundless, porous, and fragmented) recentres debate on the design of the city around the theme of the network to overcome the discontinuities that have been created within it, and with surrounding contexts (natural and socio-economic). Infrastructural, transportation, environmental, and ecological networks accompany those of historical heritage to define the set of cultural values and identities that characterise the contemporary territories. Working within the logic of the network allows us to overcome the dichotomy between reconstruction and prevention, as well as the difficulty in differentiating and specialising the range of performance of the single elements of which it is composed.

The construction of a frame comprised of thematic, though complementary and integrated networks, determines an approach to the analysis of the city and design that inverts traditional planning and design methods. Indeed, the complexity of urban systems perhaps resides much more in how they are represented than in the nature of the systems themselves. Complexity is a quality of the observing system (urban planning) and not only of the observed system (the city). Therefore, the more urbanism employs structural networks in the design of cities and territories, the more it spreads and relaunches their specific characteristics outside the elements of the network. The design of networks redistributes weights and responsibilities and generates new values and orderings of the features that make up the network. In this manner, recognised and planned networks converge toward an organised system (the frame) that structures successive grafts and changes, as the network is by definition an open, as opposed to a closed system.

The networks that intersect and complete one another define the frame inside which urban regeneration takes form. By intersecting thousands of threads (relations between nodes), the work of the frame produces a visible (feasible) image of the future city, whose purpose is threefold. This design: (i) serves to support the development of the city, whose networks (environmental, infrastructural, and historical) are no longer a backdrop against which to activate transformative processes but move into the foreground and become the condition that initiates transformative processes; (ii) serves to orient the development of single and specific interventions that find coherence at the territorial scale (of the network) despite being activated over lengthy periods, and in different places and ways; (iii) serves to order the priority of actions, to programme and hierarchies the transformations of the territory based on the maturity of projects, on the agreement reached among stakeholders and available economic resources.

Urban regeneration based on the design of a frame of networks is not about identifying the optimum solution (one best way) but one among many possible solutions in a given spatial context at a particular moment in time. Through this approach, the network (environmental, infrastructural, or historical) becomes a strategy for exploration that does not reduce complexity but articulates and decomposes it to identify the unit of intervention that can be confronted simply and concretely, while respecting the dialectic with the other components of the network, as well as the frame. The environment and history are themes/factors that most easily define the concept of the network for the regeneration of the contemporary city.

Ecological networks represent a fundamental theme for urban planning and design approached from an environmental standpoint. Green and Blue Infrastructures now envelop/amplify/develop ecological networks and the more complex system of urban, peri-urban, agricultural and semi-natural open spaces. The city of the future must be capable of absorbing disturbances and changes to climate, the environment and health, incorporating the concept of resilience, and the design of blue-green networks integrated with grey networks (underground utilities) and for mobility become strategic for supporting concrete processes of urban regeneration. "Vegetal and water networks, agrarian urban and peri-urban landscapes, leftover and wasted areas interact increasingly more often with the traditional public spaces of the street and square, qualifying them by bringing advanced ecosystemic and technological elements and penetrating into urban fabrics. They thus stimulate a paradigm shift in the urban metabolism founded on the recycling of resources and asocial and identity-based re-appropriation of common goods. Blue and green infrastructures thus become a dynamic constellation of ecologically and socially informed tactics that interact with systemic choices of a reticular nature, oscillating between synergies and conflicts and outlining place-based strategies of urban regeneration" (Gasparrini 2017).

As with open green spaces, the space of history (prevalently built, but not only) represent an interesting element atop which to build a network for the regeneration of the contemporary city. As presented in previous paragraphs, the topic of urban historicity is very dear to Italian urbanism, and has permitted the sedimentation of analytical-interpretative and design-based paradigms (cf. historic centre) and approaches to conservation, safeguarding, preservation and restoration of historical heritage (cf. a Muratorian typo-morphological approach[4]) exported at the international level. This study and continuous evolution of the discipline around the theme of urban historicity and its role to collective society now recognises the existence of a structuring character for the design of the city, considering the values of history a true infrastructure of the territory (Bonfantini 2013). As an infrastructure—a term that generally refers to complementary works necessary to economic activity (streets, railways, airports, etc.) or indispensable to new urban settlements (sewers, parks, gardens, etc.)—the set of historical values become fundamental to the design of the city of tomorrow. This is because the usefulness of the historical network derives from the benefits enjoyed by communities who make use of it and care for it. Caring for the territory, in other words, its daily and systematic comprehension, protection and promotion, becomes an action of designing for the future, and not simply a remedy for the wounds inflicted by the global changes taking place (banalisation, touristification, abandonment, degradation, etc.). Therefore, its reconstruction is essential if we wish to inject a new quality into contemporary urban spaces, re-reading meaning and qualities per contemporary society.

The multi-scale and multi-value dimension of the historical territory—which overcomes administrative boundaries and the theme of restriction—also leads to a rethinking of the traditional historical city. The historic city is intended here as that system of urban fabrics, both ancient and new, in other words, historic centres, selected nineteenth-century expansions (for example those designed by Beruto in Milan), and the architecture and districts of the Modern Movement (Campos Venuti 2008). This definition must be adapted to various urban realities whose genesis, morphology, scale, state of conservation, and socio-economic role can vary widely. In any case, the urban structure of these historicised urban fabrics is characterised by a system of relational public spaces (streets) that play a crucial role in recognising the typo-morphological and functional variants of urban history. The street is intended here as the 'place' for experiencing the space 'between buildings,' consolidating collective memories and identities and stimulating new methods of using urban fabrics. This is because, as Louis Kahn wrote, "The street is a room by agreement, a community room the walls of which belong to the donors [...]. Its ceiling is the sky" (Bonaiti 2002). Working with the street means bringing urban planning policies and actions back to the human scale and, in so doing, (re)discovering proximity as an ideal dimension for improving quality of life.

---

[4] Urban morphology is the study of city forms, while building typology is the study of building types. In the 1950s, Saverio Muratori invented the "morphogenesis of urban space" approach by identifying 'tissues' as settlement morphologies that characterise the form and spaces between streets and buildings (typo-morphological analysis). "Morphologically homogeneous urban parts can be distinguished into 'tissues and open forms' (*tessuti e forme aperte*). Tissues are the settlement morphologies characterised by a close correlation between the shape of the street spaces and the set of buildings, determined by the fact that the built fronts are arranged along the edges of the streets" (Cappuccitti 2008: 289). Tissues can be classified according to two characteristics called 'structure and grain' (*impianto e grana*). The structure is the shape of a set of streets and can be distinguished in intricate, reticular, radiocentric, and organic. The grain is the fragmentation degree of the built-up space in the fabric, and the size of buildings' footprint defines it.

In a condition of dispersion and fragmentation the city requires two actions: a re-structuring of its settlement features and an infra-structuring of its open-air spaces (not only green areas). In a perspective of connection and relation between parts, the street ceases being a mere infrastructure for transport and becomes the ordering part of the historical city. The functional components of the city, streets handling flows of goods and people, are transformed into elements that re-design the urban landscape and become the space for rooting and recognising new identities.

During the pandemic in many cities, we witnessed the creation of small public squares, some as large as a single parking stall, offered to inhabitants and users of the street for various purposes: micro-spaces created primarily for encountering others, meeting and socialising. The post-pandemic city asks for more temporary, hybrid, and reversible spaces. Many experiences in tactical urbanism have been consolidated (see the project for Piazza San Luigi e Nolo in Milan) or intensified (such as the creation of new dehors for bars and restaurants). They redesign the urban space of the street, now less valued for its form than for what occurs within it (Gehl 2011). The street becomes a specific object of study and urban design with a view toward global sustainability. Attractive and qualifying uses contaminate the street to bring value to the communities it intercepts and so that its space generates recognisability, economies, environmental quality, and historical-cultural identity.

In the wake of the pandemic, numerous cities in Europe and around the globe have attempted to provide a new impulse to public space, above all in dense and historicised contexts. This is achieved through policies and projects for improving the performance of streets, viewed as both spaces of relations and flows (to be travelled by means that offer an alternative to the automobile) and a space of socialisation and proximity. History presents well-known engineering and sanitary operations from the late nineteenth century applied to road infrastructures to improve conditions of salubrity and hygiene in the city (London, Paris, Naples, Philadelphia, New York). Instead, recent experiences in re-qualifying the street consider infrastructural space for new integrated uses based on more sustainable mobility.

In various projects proposed in different cities—Paris (Plan Velò, 2015), Auckland/New Zealand (Innovating Streets for People, 2016), Barcelona (Eixample Superilles, 2016), Milan (Strade Aperte, 2018), Montréal (Pedestrian Only Streets, 2020), Mountain View/USA (Castro StrEATs Summer, 2020), New York (Open Streets, 2020), Rotterdam (Witte de Withstraat, 2020), Turin (Precollinear Park, 2020)—the street is not merely a traffic artery but a space of social aggregation. The total or partial closure of streets, reduced lane widths or the elimination of parking stalls are accompanied by the introduction of functions for sport, leisure, play, culture (exhibitions, concerts) and other services, promoting both pedestrianisation and cyclability, as well as the street space as a place of pause (Guzzabocca and Legoratti 2021).

In most cases, these operations are carried out within existing historical fabrics that developed after the Industrial Revolution. A city with vast road infrastructures was once dedicated to the 'rapid' passage of carriages, but also to favour strolling by the bourgeoisie under the shade of large trees. Today's *controviali* (frontage roads) were, in reality, broad sidewalks, along which Parisians, Turinese, and Milanese strolled and stopped to enjoy a coffee in ground floor shops fronting the street. These broad systems for circulation and social life in the city were reinterpreted in programmes to pedestrianize parts of the city, which only intensified with the pandemic and allow residents to experience a new urbanity.

Similar interventions demonstrate how the re-activation of the public space of the street consolidates the presence of communities in the neighbourhoods of historical cities. Returning people to the street reaffirms the necessity to establish a relationship between space and society that consolidates urban identities and attributes meaning and significance to places.

3.3 Slow Tourism and Proximity also in the Contemporary City

3.3.1 New Itineraries into the Historical Territory

The theme of the historic city is inevitably linked to the question of its use for tourism. Monuments, historic centres, modern architecture and districts are a heritage whose historical, artistic, aesthetic, cultural, and recreational value and use can be appreciated by inhabitants and users of the city alike. When the theme of proximity is linked with identity in policies for rebalancing the city, then the proportion between conservation and innovation acquires a strategic role in redefining the space in which communities identify themselves, without bending to exogenous forms of use. New bonds with the territory are evidence of a 'plural city' with a growing ethnic, cultural and religious variety; on the other hand, the historic city also becomes a privileged space for (global) tourism as a concentration of collective life experience and memory. The historic city is the principal expression of the art of organising urban space. Its strong representativeness is considered a resource for attracting tourism.

In recent decades, strong criticisms have been moved against mass tourism or overtourism (Koens et al. 2018). This approach has created problems for the liveability and identity of many large cities, such as Venice, Florence, and Rome, and labelled cities of art as fun parks. Thanks to heightened accessibility to low-cost flights, to new digital platforms for the autonomous organisation of travel, not to mention the influence of global organisations like UNESCO, responsible for promoting a series of 'must-see tourist destinations,' historical cities have developed forms of voracious and hurried fruition. The UNESCO label is not a cause of tourism, but its certification, its guarantee, be-cause it produces destructive effects on the preservation of cultural heritage at all costs (D'Eramo 2017). Mass tourism, also known as 'hit-and-run' tourism, is a way of using the city. Its public spaces are crowded with tourists generating difficult conditions for residents (simply strolling in the centre has become stressful and laborious in many cities). Moreover, mass tourism creates problems in terms of management of services, both public (waste collection, mobility and public transport) and private (the offering of lodgings and restaurants).

The 'science of tourism' and international bodies have been looking for some time at a more curious, more sustainable and responsible form of tourism (cf. the website of the World Tourism Organization https://www.unwto.org/sustainable-dev elopment). Sustainability has become central to promoting conditions of use with a lesser impact on society and the environment. In recent years, particularly in the wake of the pandemic, there has been a growing affirmation of the phenomenon of under-tourism. It is a practice of traveling to discover unusual and lesser-known destinations, often national and regional, and for this reason, less crowded (Mihalic 2017). Cities around the world are working to appeal to the collective imagination through campaigns of marketing designed to promote less famous destinations in a bid to 'save' traditional sites from over-tourism.

As Zygmunt Bauman reminds us, contemporary society's current freedom of movement has become a 'must-have' that is difficult to do without. However, it also becomes a criterion of accessibility to spaces and services, places and functions, attractions and sources of enjoyment. "The effects induced by the new condition create radical inequalities. Some of us become 'global' in the true and fullest sense of the term, others remain fixed to their 'localness'—a condition that is anything but pleasing or supportable, in a world in which the 'globals' set the rhythm and establish the rules of the game of life" (Bauman 2001: 5).

The challenge presented to planners and decision-makers is thus focused on defining urban strategies capable of stimulating new flows of tourism—more sensitive and aware, but also more sophisticated, connected, and emancipated—capable of promoting visits to (global) 'must-see destinations' without denaturing their structures and identities (Coliva 2021). The search for a balance between crowds of tourists and empty streets or museums is preferred, and some believe this objective can be reached through 'creative tourism,' which employs existing or potential resources (Gowreesunkar and Vo Thanh 2020). In this perspective, it is crucial to work with the performance of medium- to large-sized cities, otherwise excluded from circuits of 'slow tourism,' which tends to search primarily for unknown 'villages' and natural settings. In particular, it becomes strategic to work with an existing network of historical territories related to the needs of residents and tourists, offering extensive spaces, services, and uses that are both original and sustainable.

In fact, tourist and historical networks can be combined, expanding occasional fruition to different parts of the city, even in large cities. The historic city and its heterogeneous public spaces can be connected and enjoyed by providing new routes in many ways. By offering itineraries, these new routes can bring people closer to discovering historical and natural heritage (ecotourism, geotourism), and become familiar with proximity, i.e. the legacies found close to home (staycation, locavism, according to Hollenhorst et al. 2014) and unveil neighbourhood identities. Moreover, routes can also be new in terms of the way they are travelled (biketourism), and how they involve tourists in visiting (material and immaterial) heritage (voluntourism).

The development of slow tourism and staycationing explains the interest of cities and urban planning in creating paths, circuits, and itineraries for rediscovering the stratification of signs, memories, and new identities, inside and outside the dense city (Imbesi 2003). Approaching sustainable and responsible tourism in the wake of the pandemic signifies offering spaces and services to travellers looking to 'experience place,' supporting the competitivity of local businesses, and promoting territorial excellence (ancient or modern).

Regarding consumer and mass tourism, new vacationers are looking for different forms of travel concerning the past (bike-tourism and public transport), preferably outdoor and immersed in nature (rural and enogastronomic tourism). In the long-term, these requests may pose a risk to the economies of many large cities and traditional destinations. To avoid dying out, they will have to equip themselves with new infrastructures more suited to the changing needs of society and tourism focused on authenticity of place.

Big cities can thus rediscover the fabrics of the historic city and the identities of diverse districts, promoting slow itineraries through urban streets that reveal the rich legacy of signs. Itineraries that combine tourism circuits and pedestrian, bicycle, or waterborne routes and that, in addition to being more sustainable, also help present and appreciate unusual or lesser-known parts of the city (innovative industrial districts, Modernist buildings, working-class neighbourhoods). Itineraries that simultaneously aid and decrease crowding in traditional tourist sites (historic centres, museums, cathedrals, monuments) and increase the quality of environments and lifestyles for residents (reduced traffic, accidents, noise, pollution, and increased local services).

For tourism to be reconfigured toward a more creative and sustainable dimension, cities must also contribute to supporting it by reimagining their offering of facilities, public spaces and connections. Both must pursue the objective of overcoming the myopic vision of promoting monumental historical-cultural heritage as a means for generating profit (to be used for successive investments) and instead promote the historical territory as the means for maintaining and caring for the authenticity of place over time.

Globalisation, supported by technology, social media, and low-cost travel, has stimulated the interest in discovering both unusual and traditional destinations that have been capable of renewal through urban projects and works of contemporary architecture. Paris (the Beaubourg, the Museo d'Orsay by Gae Aulenti, the Pyramid by Ieoh Ming Pei, the Défense Grand Arche, the restoration of the Fondation Cartier pour l'Art Contemporain by Jean Nouvel, the new Fondation Louis Vuitton by Frank Gehry in the Jardin d'Acclimatation inside the Bois de Boulogne), London (the Millennium Wheel, the renovated spaces of the Tate Gallery and the Dome along the River Thames), Berlin (Renzo Piano's Potsdamer Platz, the Holocaust Memorial by Peter Eisenman), Bilbao (the Guggenheim Museum by Frank Gehry and the white Zubi Zuri footbridge by Santiago Calatrava) are the traditional European tourist destinations that have taken on a new life for many young and less young travellers, attracted by the discovery of new values. American and Asian metropolises join these European cities: from Seattle (home to the historical Space Needle and the Experience Music Project by Frank Gehry), to Los Angeles (the Getty Center by Richard Meier) to Dubai, the global capital of tax-free shopping and land of experimentation with audacious works of architecture and new districts for tourism (Pascucci 2015). In this condition of globalised and globalising tourism, emphasising uniqueness and defending historical heritage-identity can be a tool for the survival of traditional cities.

The network of historical territories permits a re-reading of historical heritage, in order that it can be protected and, at the same time, promoted. This expands the boundaries and dilates the experiences of fruition; it enriches it with cultural values. The design of the network of historical territories—which defines new relations, invites further readings and understandings and critically reformed observation—contributes to determining the infrastructure of the city, which produces new images and new sources of attraction. The design of the historical network preserves the spirit and characteristics of valuable resources, investing in factors with a lasting and important attractiveness. This project dilutes the attractiveness of historical-cultural heritage across a broad spectrum of resources. On the one hand, it facilitates the arrival and permanence of mass tourism because it is captured by a new, richer, articulated and unprecedented cultural offering. On the other hand, it helps distribute tourists across different spaces and areas, keeping people from concentrating in the usual parts of the city, which causes congestion.

Cities can thus redefine new itineraries for the fruition of historical territories, reinvigorating their role as sites of collectivity, in which architecture (from the ancient to the modern) assumes a socially communicative and interactive function for its ability make urban space a place in which people can recognise and identify themselves. In this sense, the redesign of particular 'urban thresholds,' areas of passage and gateways to the city for flows at a vast scale, may help expand the range of places whose identity is a bearer of historical interest. For example, the project for the renewal of Italy's main railway stations (from the Central Station in Milan to Termini Station in Rome) represents nodes in a frame composed of different networks: from the infrastructural (at the vast scale) to the historical. These spaces, once requalified, can obtain the same appreciation as the heritage of the past; they convert railway stations into dynamic public spaces and an area for exchanges between different cultures. Additionally, when these urban thresholds are connected and inserted within a circuit of fruition based on soft, active and sustainable mobility, it is possible to further expand the network of historical territories, distributing it and linking it up through urban and extra-urban itineraries with slow, sustainable, unique and greener circuits.

3.3.2 The Continuous and Daily Care for Heritage

As Rantala and her colleagues write (2020), with respect to the definition of a new model of di tourism after the Anthropocene more connected with the nature of places, Rhythmicity—the revealing of cycles and biorhythms shared by different people, more than a simple coexistence among them—Vitality—the self-organisation of living beings to coexist and positively influence one another, unhinging hierarchies of species—and Care—the existence of multiple forms of relations based on ethics (good/bad) and not on morals (right/wrong)—are three aspects with which to work in order to become reactive toward what we encounter as we discover new places. It is a question of defining emancipated forms of tourism of proximity, more sustainable and informed, and which therefore avoid the contradictions of a model of tourism camouflaged behind adjectives of green and slow (which use online platforms to book stays or low-cost and energy consuming flights), but which are truly oriented toward the promotion of quality (of places) rather than quantity (of tourists) (Izcara Conde and Cañada Mullor 2020).

In these terms, it is fundamental to work above all with the available offering of historical and cultural heritage and identity for use by tourism. As mentioned, a heritage is variegated in its form, era and state of conservation, dimension, and location. A collection of heritage so vast that it must be constantly monitored. Day-to-day care of this heritage must be at the base of living and dwelling, that is, the greatest revolution for unhinging lifestyles and models of development that threaten man and the environment in which he lives. It is not enough to physically live in a given territory to be a true inhabitant. Inhabitation means interacting and rediscovering a sense of belonging to places. We truly 'inhabit' when streets, squares, gardens, parks and buildings and every other part of the territory in which we are 'guests' (another reason why heritage represents a gift to be passed on to others and of which we are only temporary custodians), stimulate a desire to know more about and care for them.

The landscape and heritage are often private property conceded for public use (whose fruition is never exclusive but collective) whose true purpose is to satisfy people's fundamental rights. Hence they are private property (above all works of modern architecture) that contribute to the creation of a common good (the beauty and quality of the urban landscape and environments of life). In this perspective of collective function, the preservation and safeguarding of historical heritage are not objectives to be pursued, but the means through which to amplify and spread a new awareness, a new education and a new culture. Historical heritage is thus assigned an edifying role that, also supported by technologies and contemporary forms of exchange of information (high-speed and for the masses), spreads and expands accessibility to the resources offered by the territory (Montanari 2017). Resources that, to remain structuring of the identity of communities, and thus structural elements to the definition of the identity of places, must be constantly cared for.

Of Latin origin, the word 'care' is often utilised to indicate affective relations and in reference to an attitude of thoughtfulness. In care there is an action of paying attention and supporting, in the broadest sense of cultivating a relationship or (indicated by the derivation of the term 'culture') of building, constructing and producing awareness.

Pragmatically, an 'inhabitant' (resident, tourist, and worker) who wishes to care for existing historical heritage to pass it on as a source of knowledge and testimonial has three essential tasks: to be a good custodian, to become a good observe and gain awareness of his/her limits of action. Being a good custodian means paying attention to heritage (be it mobile or immobile), knowing the correct procedures to be implemented to ensure its conservation and durability over time and being a valid support to those legally responsible but unable to do this with the necessary consistency. Knowing how to observe means being able to capture those signs that may suggest deterioration or a potential source of risk to the conservation of heritage. Acting with awareness of one's limits means avoiding improvised restoration or conservation works and promoting regular cleaning and ordinary maintenance of heritage and the context in which it is situated.

3.4   Defining a Theme for Each Station of the New M4 Metro Line

The experience of action-research for the design of the Green and Blue Backbone of the new M4 metro line is important to the re-signification of places and for rooting the identities of neighbourhoods. The historic city of Milan, the urban result of the 'spontaneous' evolution of the ancient centre and the 'planned' evolution of nineteenth/twentieth-century fabrics (Oliva 2002), is now placed in a condition of tension by the construction of the city's fifth metro line. The realisation of the M4 was welcomed as an opportunity to restore meaning and significance to the sites it intercepts and comes into contact with through its stations. When designing this Backbone, the 21 stations of the new metro line were considered the nodes of a single Green and Blue Infrastructure for the city that links (subterranean) mass public transport, soft mobility, waterways, green systems, history, and local identities. Within this vision of infrastructure at the service of residents and city users, the Green and Blue Backbone becomes an occasion for bringing new meaning to places in the city by thematising the stations based on the characteristics of the neighbourhoods in which they surface.

Metro Art Los Angeles and the Naples metro (both completed during the 1990s) are well-known examples of how the thematisation of transport stations can permit the construction of true and proper stations-landmarks for their respective cities. In these examples, the arrival of the metro and the construction of its stations in different neighbourhoods brought an identity to the 'generic city' (Koolhaas 1997). Working in the opposite direction, with the Green and Blue Backbone project, the thematisation of the metro stations occurs by recognising the peculiarities (environmental, historical, and functional) of the territory in which the infrastructure emerges on the surface. The design of the stations at the level of the city is not focused on the theatricalization, musealisation or artificialisation of Milan's public space—using an artistic sign (installations, artworks, video, colours, etc.)—, tagging the stations and, with them, the neighbourhoods in which they are located. Instead, the objective is to highlight the inherent resources offered by the diverse territorial contexts traversed by the M4 line and reveal the (already existing) value of urban space. In other words, it is not the station that thematises place, but the site (with its services, history, monuments, typo-morphological layout, ecological devices and identity) that attributes a design theme to the station (Fior 2020a; b).

This approach to design orients future architectural projects for the (prevalently public) spaces around the stations, working with the identity of place and proximity to the community. The objective is to stimulate the day-to-day care of regenerated spaces, permitting the historic city revealed by the project (façades of basilicas, archaeological remains, ancient watercourses, Modernist Milanese architecture) to be constantly surveilled and maintained by communities. As part of the project for the Green and Blue backbone, the historic city, whose sense and meaning are re-actualised, acquires the role of a driver for the regeneration of the settlement system and loses the connotation of something subject to a passive heritage listing to ensure its safeguarding and protection (Figs. 3.1, 3.2, 3.3, 3.4 and 3.5).

References

ANCSA—Associazione Nazionale Centri Storico-Artistici (1990) Carta di Gubbio 1990. Retrieved from https://www.ancsa.org/la-storia-e-larchivio/la-seconda-carta-di-gubbio-1990/

Bonaiti M (2002) Architettura è. Louis I. Kahn, gli scritti, Mondadori Electa, Milan

Bonfantini B (2002) Progetto urbanistico e città esistente: gli strumenti discreti della regolazione. Libreria CLUP, Milan

Bonfantini B (2013) Centri storici: infrastrutture per l'urbanità contemporanea. Territorio 64:153–161. Franco Angeli, Milan

Bonomi A, Abruzzese A (eds) (2004) La città infinita. Mondadori, Milan

Campos Venuti G (2001) Il sistema della mobilità. In: Ricci L (ed) Il nuovo piano di Roma. Urbanistica 116:166–172. INU Edizioni, Rome

Campos Venuti G (2008) La città storica tra passato e futuro. In: Evangelisti F, Orlandi P, Piccinini M (eds) La città storica contemporanea. Edisai, Bologna, pp 88–93

Cappuccitti A (2008) Tessuto urbano. In: Mattogno C (ed) Ventuno parole per l'urbanistica. Carocci, Rome, pp 289–295

Cellamare C (2009) Processi di costruzione delle identità urbane: pratiche, progetto, senso dei luoghi. Geotema 37:75–83

Cervellati PL (2010) Centri Storici. Treccani. Retrieved from https://www.treccani.it/enciclopedia/centri-storici_%28XXI-Secolo%29/

Choay F (1973) La città. Utopie e realtà, vols 1 and 2. Einaudi, Turin

Choay F (1992) L'orizzonte del post urbano. Officina, Rome

Choay F (1995) L'allegoria del patrimonio. Officina, Rome

Choay F (2008) Del destino della città. In: Magnaghi A (ed) Alinea Editrice, Florence

Coliva A (2021) Una nuova cultura del turismo per le città d'arte. Il Sole24ore, 6 Feb 2021. Retrieved from https://www.ilsole24ore.com/art/una-nuova-cultura-turismo-le-citta-d-arte-ADoZzmHB

Corboz A (1985) Le territoire comme palimpseste. Diogène, no. 121, It. tran., Il territorio come palinsesto. Casabella 516:22–27. Mondadori, Milan

D'Eramo M (2017) Il selfie del mondo. Indagine sull'età del turismo. Feltrinelli, Milan

Di Biagi P (1999) I quartieri: 'patrimonio' del moderno. Urbanistica Informazioni 168:5. INU Edizioni, Rome

EC—European Commission (2020) The new leipzig charter. The transformative power of cities for the common good, Adopted at the Informal Ministerial Meeting on Urban Matters on 30 November 2020. Retrieved from https://ec.europa.eu/regional_policy/sources/docgener/brochure/new_leipzig_charter/new_leipzig_charter_en.pdf

Fior M (2013) I territori storici della città contemporanea. PhD Thesis XXV cycle in Government and territorial design, Politecnico di Milano, Milan

Fior M (2020a) Milano. Il progetto della Dorsale verde-blu: tessuti storici e città pubblica. In: Poli I.: Città esistente e rigenerazione urbana. Per una integrazione tra Urbs e Civitas. Aracne Editrice, Rome, pp 154–169

Fior M (2020b) The history of urban conservation policy for historic cities in Italy. In: Inoue N, Orioli V (eds) Bologna and Kanazawa. Protection and valorization of two historic cities. Bononia University Press, Bologna

Gabellini P (2010) Fare urbanistica. Esperienze, comunicazione, memoria. Carocci Editore, Rome

Gabrielli B (1993) Il recupero della città esistente. Saggi 1968–1992, Etas s.r.l., Milan

Gasparrini C (2008) Strategie, regole e progetti per la Città storica. Urbanistica 116:93–105. INU Edizioni, Rome

Gasparrini C (2017) Le infrastrutture verdi e blu nel progetto della città contemporanea. Introduction to workshop organized by INU at 'Biennale dello Spazio Pubblico', 10 April 2017, Rome

Gehl J (2011) Life between buildings: using public space, 1st ed. 1971. Island Press

Gowreesunkar VG, Vo Thanh T (2020) Between overtourism and under-tourism: impacts, implications, and probable solutions. In: Séraphin H, Gladkikh T, Vo Thanh T (eds) Overtourism: causes, implications and solutions. Palgrave Macmillan, Cham, pp 45–68. https://doi.org/10.1007/978-3-030-42458-9_4

Guermandi MP, D'Angelo U (eds) (2019) Il diritto alla città storica. Atti del Convegno—Roma, 12 November 2018, Associazione Ranuccio Bianchi Bandinelli

Guzzabocca C, Legoratti S (2021) Una nuova prossimità. La riappropriazione della strada ai tempi della pandemia. Master's Thesis in Progettazione dell'Architettura, Politecnico di Milano, A.Y. 2020–2021, Supervisor Prof. M. Buffoli, Co-Supervisor Prof. M. Fior, Milan

Halbwachs M (1987) La memoria collettiva, 1st ed., It. tran., Edizioni Unicopli, Milan

Hollenhorst SJ, MacKenzie SH, Ostergren DM (2014) The trouble with tourism. Tour Recreat Res 39:305–319. https://doi.org/10.1080/02508281.2014.11087003

Helzel PB (2016) Il fondamento dell'identità nella dialettica tra memoria e ricordo. Bollettino Filosofico 31:176–194. https://doi.org/10.6093/1593-7178/4048

Imbesi NP (2003) L'effetto pulsar. Azienda Turismo, Anno V—Nuova Serie, no. 1/2, Septtemer

Izcara Conde C, Cañada Mullor E (2020) Slow tourism, una oportunitat per a la transformació del turisme? Tour Herit J 2:110–122 (2020). https://doi.org/10.1344/THJ.2020.2.8

Koens K, Postma A, Papp B (2018) Is overtourism overused? Understanding the impact of tourism in a city context. Sustainability 10:4384. https://doi.org/10.3390/su10124384

Koolhaas R (1997) Generic city. In: Koolhaas R, Mau B (eds) S, M, L, XL, It. tran., La città generic. Domus 791. Mondadori, Milan

Magnaghi A (2000) Il progetto locale. Verso la coscienza di luogo. Bollati Boringhieri, Turin

Manieri Elia M (2008) La Città storica struttura identificante. Urbanistica 116:109–116. INU Edizioni, Rome

Mazzoleni C (1991) Dalla salvaguardia del centro storico alla riqualificazione della città esistente. Trent'anni di dibattito dell'ANCSA. Archivio di Studi Urbani e Regionali 40:7–42. Franco Angeli, Milan

Mihalic T (2017) Redesigning tourism in CEE countries: the main areas of change and the communist past. Int J Tourism Cities 3(3):227–242

Montanari T (2017) Il paesaggio e il patrimonio storico e artistico: un unico bene comune. Le Nuove Disuguaglianze. Beni Comuni 2:66–71

Musco F (2016) Nuove pratiche di rigenerazione in Europa. In: D'Onofrio R, Talia M (eds) La rigenerazione urbana alla prova. Franco Angeli, Milan, pp 49–63

Oliva F (2001) Il sistema ambientale. In: Ricci L (ed) Il Nuovo Piano di Roma. Urbanistica 116:158–165. INU Edizioni, Rome

Oliva F (2002) L'urbanistica di Milano. Quel che resta dei piani urbanistici nella crescita e nella trasformazione della città. Hoepli, Milan

Oliva F (ed) (2010) Città senza cultura. Intervista sull'urbanistica a Giuseppe Campos Venuti, Laterza, Rome-Bari

Pascucci D (2015) Architettura di Qualità e Attrazione Turistica. Academy Formazione Turismo, December (2015). Retrieved https://academy.formazioneturismo.com/architettura-di-qualita-e-attrazione-turistica/

Rantala O, Salmela T, Valtonen A, Höckert E (2020) Envisioning tourism and proximity after the anthropocene. Sustainability 12:3948. https://doi.org/10.3390/su12103948

Ricci L, Battisti A, Cristallo V, Ravagnan C (eds) (2018) Costruire lo spazio pubblico. Tra storia, cultura e natura. Urbanistica DOSSIER, 15. INU Edizioni, Rome

Rossi A (1995) L'architettura della città, 2nd edn. CittàStudiEdizioni, Turin

Secchi B, Viganò P (1998) Piani e progetti recenti di studio 1998. Un programma per l'urbanistica. Urbanistica 111:64–76. INU Edizioni, Rome

Simmel G (1908) Sociologie. Untersuchungen über die Formen der Vergesellschaftung, It. tran. 1998, Sociologia, Edizioni di Comunità, Turin

Talia M (2017) Un futuro affidabile per le città. Planum Publisher, Rome-Milan

Toppetti F (ed) (2011) Paesaggi e città storica. Teorie e politiche del progetto. Alinea Editrice, Florence

Volpiano M (ed) (2011) Territorio storico e paesaggio. Conservazione Progetto Gestione. L'Artistica Editrice, Savigliano (CN)

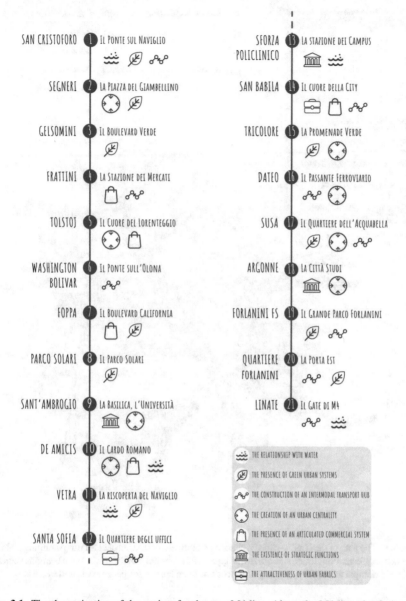

Fig. 3.1 The thematization of the project for the new M4 line. Along the M4 line, the design of areas on the surface of the stations (left) is guided by a theme (right), in turn defined by a set of themes and questions (the symbols) that emerged from a critical-interpretative study of the urban setting

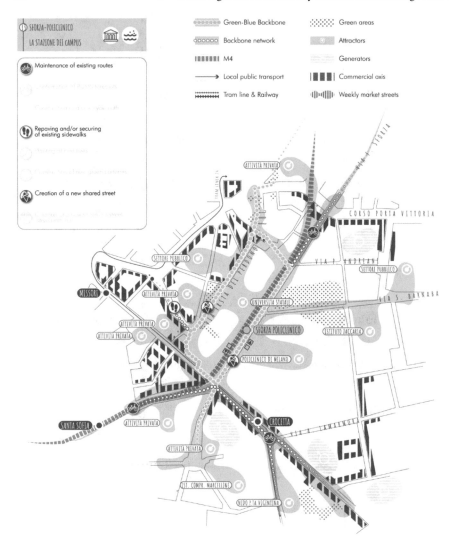

Fig. 3.2 The path of the Green–Blue Backbone and the relative masterplan for programming actions around the Sforza-Policlinico station. In the upper left is the theme of the project (the Campus station) and the principal actions for creating a network of open spaces for pedestrians and cyclists

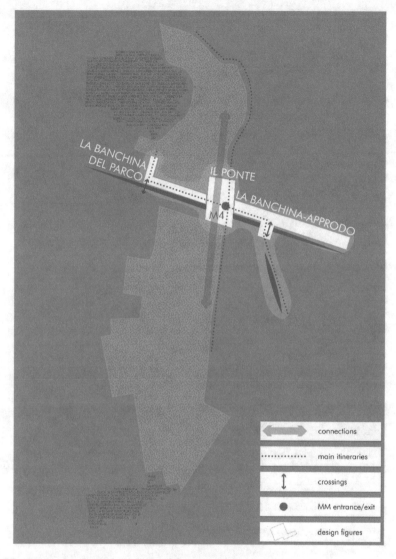

Fig. 3.3 The 'design figures' for Vetra station. The diagram highlights the places generated by the redesign of the surface areas near the M4 entrances/exits. In particular, the project for Vetra station includes a connection between two green areas of the Parco delle Basiliche (the bridge); and the development of a system of open spaces for pedestrians (the park quay), coherent with a renewed urban landscape that rediscovers the city's historical heritage and new forms of tourism (the quay-mooring point along the Naviglio)

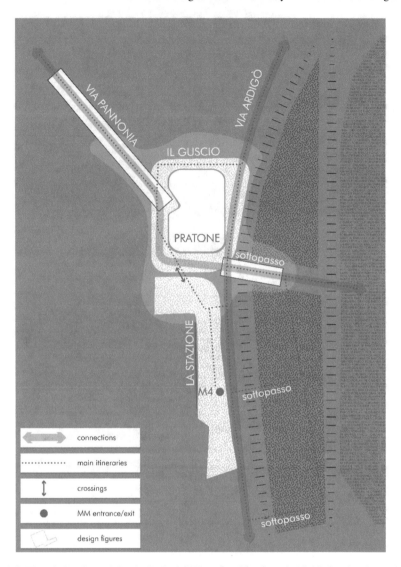

Fig. 3.4 The 'design figures' for the Forlanini FS station. The diagram highlights the places generated by the redesign of the surface areas near the M4 entrances/exits. In particular, the project for the Forlanini FS station proposes two specific areas: 'the lawn', a usable green space at the heart of the project; 'the shell', trees and paths creating a buffer area that protects the heart of the lawn; the urban axis of 'Via Pannonia' to be rehabilitated and reorganised by introducing bicycle-footpaths connecting the station with the city; 'the underpass', a project to reuse an underpass once used to transport building site debris to connect the city's cycle-footpaths with metropolitan networks; and, finally, the area of the 'station', characterised by a redesign of permeable and filtering surfaces connected with the proposed introduction of new soft mobility itineraries

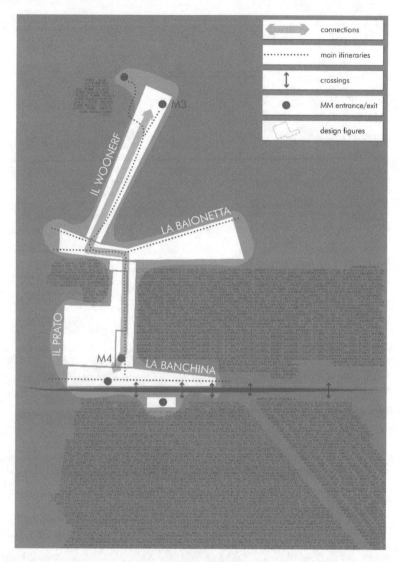

Fig. 3.5 The 'design figures' for the Sforza-Policlinico station. The diagram highlights the places generated by the redesign of the surface areas near the M4 entrances/exits. In particular, the project for the Sforza-Policlinico station identifies four project areas. Most importantly, 'the quay' where the design of public space links the insertion of the exits from the M4 with the future presence of the Naviglio and creates a space of connection between the university campus and the hospital; 'the lawn', a very large area that includes both the redesign of the historical tree-lined pedestrian path along the edge of the State University as well as the areas belonging to the Basilica di San Nazaro in Brolo; the area of Largo Richini (the bayonet) transformed from parking into a public square; and, finally, the axis of Via Pantano (the woonerf) whose requalification and pedestrianisation will create a space of connection between the M3 and M4 lines

Chapter 4
Safety, Green and Blue Networks, Active Mobility and Walkability

Abstract In this chapter, proximity is defined in relation to urban mobility and ecological connections, with infrastructures and the environment representing two essential networks for the regeneration of contemporary cities. The fragility of cities (territorial vulnerabilities and social divisions) imposes a rediscovery of the notion of caring and the wellbeing of residents (health and safety). This is possible by promoting paths for slow and active mobility, completed by and unfolding within vaster ecological-environmental and landscape corridors. The quarter becomes a field for experimenting with new variable geometry networks, to be calibrated concerning needs of accessibility and inclusion, of public health, not to mention the implementation of an ecological and digital transition. In this vision, the base units of settlement resemble the cells of a healthier body. The action of caring becomes an attitude toward design focused on increasing resilience toward future challenges, rather than a solution to the problems of the past.

The chapter develops three interconnected topics. The first is linked to the unsustainability of the contemporary city, founded prevalently on vehicular mobility, to be corrected by introducing forms of movement that are less energy-consuming, more diversified (integrating the so-called *cura del ferro*, the improvement of rail-based transport) and more socially inclusive. Incorporated within an urban mobility project this theme translates, on the one hand, into the preparation of a plurality of mobility infrastructures that reduce the vulnerability of the urban system in the face of possible critical events (overcrowding of public transport, mandatory social distancing, and roadworks). On the other hand, it means paying attention to people with limited psychomotor capabilities and society's more delicate groups, such as children and the elderly. Successively, the chapter looks at the theme of green and blue networks that in literature assume the role of a territorial infrastructure for environmental and ecosystemic connections at the vast scale, though which can be difficult to implement at the urban scale in dense and historicised contexts. This topic reflects on the feasibility and methods of implementing an actual ecological urban transition. Finally, the third topic looks at active mobility in the city, in other words, the need to offer infrastructures that ensure connections (also longer than 15 min) designed to stimulate the physical movement of people (including many elderly people, or those with chronic pathologies or sedentary lifestyles), working toward a dimension of proximity linked also to new directions in the territorial reorganisation of healthcare. The chapter concludes with the presentation of various solutions proposed in Milan with the Green and Blue Backbone project.

4.1 Pedestrianisation and Innovative Mobility to Reduce Urban Traffic

4.1.1 Unsustainability and Multi-Risk Age

The Anthropocene is the era of pandemics. The current planetary diffusion of the SARS-CoV2 virus, responsible for the death of more than 4 million people since early 2020, is the most serious 'biological pandemic' ever faced by contemporary society (source: Italian Ministry of Health, July 2021).[1] Climate change, caused by the exponential rise in atmospheric temperature, has been a true 'environmental pandemic' that causes some 5 million deaths a year (data provided by a group of Chinese researchers at Monash University and Shandong University, recently published in the review *The Lancet Planetary Health*). Finally, sedentariness—resulting from physical inactivity, contributes to problems of obesity, cardiovascular illnesses and risk of a heart attack, in addition to aggravating illnesses linked to depression, stress and anxiety (Kohl et al. 2012)—and is considered the most difficult 'health pandemic' to subvert.

The current geological era, during which the terrestrial environment is strongly conditioned at the global scale by human action, sets us before an evident truth: humankind's ability for self-destruction and ability to cause, through social and economic choices, millions of deaths by its own hand. This attitude affects advanced and developing nations alike, obviously with widely different abilities to react, demonstrating that if the environment is fragile, so too are people.

We are what the German sociologist Ulrich Beck defined the '*risikogesell-schaft*/society of risk' (Beck 2018). The point is that, at the global scale, people and communities are in a state of constant alert due to biological, climatic, environmental, health, political and economic emergencies that succeed one another, often summed together, and reciprocally conditioning one another (in fact, we find ourselves living in the so-called 'multi-risk age'). For example, poverty and child labour have increased precisely over the two years due to armed conflicts, climate risk and the Covid-19 epidemic (WBG 2020; ILO and UNICEF 2021).

Global changes underway and this constant state of emergency are an obstacle (it would be better to imagine them as a challenge) to realising more sustainable, intelligent, safe/healthy, active and wise cities. We have been questioning the priorities to be dealt with if we wish to invert our direction for some time now. Whether we are dealing

---

[1] Covid-19—Situazione nel mondo (salute.gov.it).

with adopting a resilient or anti-fragile attitude, the objectives are first and foremost the spread of a culture of sustainability, followed by adaptation and prevention, as part of a radically renewed approach to urbanism (Gasparrini 2017; Morello 2021). If, as the philosopher Heraclitus (6th c.) wrote, the only constant in the Universe is change, as inhabitants, architects, planners and decision-makers, we must be able to condition change and not be conditioned by it. In this sense, anticipating events and making sure we are prepared to face the coming challenges is as urgent as necessary. If predicting is illusory, controlling is realistic, focusing on the best possible result in conditions of uncertainty. However, the object to be controlled is not the natural environment that surrounds us but the actions of humankind. Actions that in the long-term weigh on the environment and diminish the capacity for auto-regeneration of such fundamental resources as air, water and soil.

The pandemic was an 'urban phenomenon' that principally affected the world's large metropolises (home to over 90% of all cases[2]). However, "Density of cities has not been the decisive factor in increased infection or mortality rates of Covid-19. Rather, access to services, demographics, pre-existing health conditions, social infrastructure, and timely response measures determined the scale and impact of the pandemic" (UN-Habitat 2021: 23–24). Additionally, forecasts indicate that cities will remain powerful attractors of social and economic exchanges, making it fundamental to manage the question of over-crowding with respect to the availability of services; to improve the quality of urban spaces and the places and sectors of 'urban living,' such as, for example, the system of public transportation. "Accessible, safe, and affordable public transport is central to reducing the negative effects of spatial inequalities and segregation, both in response to the pandemic and in preparation for a secure, lasting recovery from it" (UN-Habitat 2021: 29).

4.1.2 People and Goods

The Covid-19 pandemic was perhaps the most traumatic shock recorded in recent decades in big cities for its unexpectedness, even if experts had statistically forecast its certain arrival in the wake of previous viral threats in 2003 (SARS), 2009 ('Swine Flu') and 2014 (Ebola) (Quammen 2012; Kurcharski 2017). Unfortunately, however, cities suffer sudden 'shocks' and, above all, continual 'stresses' with an equivalent effect on the quality of life. They certainly include vehicular traffic that, other than affecting the safety of residents by causing accidents (due to high speeds and too many double-parked cars), also threatens peoples' health due to polluting emissions and ambient noise.

---

[2] https://www.un.org/en/coronavirus/covid-19-urban-world.

The European Commission tells us "Every year, more than 400,000 people in the EU die prematurely due to the consequences of air pollution: this is more than 10 times the toll of road traffic accidents. Another 6.5 million people fall sick as air pollution causes diseases such as strokes, asthma and bronchitis. Air pollution also harms our natural environment, impacting both vegetation and wildlife: almost two-thirds of Europe's ecosystems are threatened by the effects of air pollution".[3] What is more, the report 'The Future of the Last-Mile Ecosystem' released by the World Economic Forum (WEF 2020), estimated that e-commerce (which grew exponentially during the pandemic) will further increase the number of vehicles on the road by some 36% in the world's largest cities by 2030, increasing CO_2 emissions by 32%. This is only one example of the unsustainability of road transport at the global scale.

At the national level, the Legambiente[4] study on the environmental performance of Italian cities shows that since 2017 Italy witnessed a progressive increase in the average area of pedestrian islands per inhabitant: from 0.45 m^2 in 2017 to 0.47 m^2 in 2018 and 2019, to 0.48 m^2 in 2020. However (unfortunately), the level of motorisation in Italian capitals continues to rise inexorably: 63.3 automobiles for every 100 inhabitants in 2017, 63.9 in 2018, 64.6 in 2019 and 65.7 automobiles for every 100 inhabitants in 2020. On this last point, the dynamics of Central Italy and the Islands are particularly worrying, as the private automobile continues to dominate. However, the number of deaths and injuries in car accidents per 1000 inhabitants (most recent data from 2019) speak to particularly critical situations in Lombardy, Tuscany, Liguria, the capital cities of Emilia Romagna and Apulia (Legambiente 2020, 2021).

The need to adapt the right to mobility to the right to a satisfying quality of life, not negatively influenced by the impacts of traffic, comports the search for a balance. A balance which can be struck through adequate urban policies and strategies, but also through integrated (mobility/city) and sustainable (energy, environment, soft mobility, and attractive public spaces) urban design (Fior et al. 2022a).

The internationally supported alternative to vehicular transport is soft mobility (pedestrian and bicycle) in synergy with more efficient and less energy-consuming public transport. Moving through the city on foot is the most natural and sustainable way to improve the quality of life in the urban environment. Indeed, it is not enough to provide ecological local transport intermodally connected with bicycle paths;

---

[3] https://ec.europa.eu/environment/air/cleaner_air/index.html#introduction.

[4] Italian environmentalist association, instituted in 1980. Legambiente is involved in awareness-raising campaigns and mobilisation to defend the territory, the environment and human health. The association combats problems of smog, nuclear energy, unauthorised construction, and illegal waste dumps. Additionally, Legambiente promotes alternative and renewable energies, energy savings and the conservation of protected areas.

instead, the cities of the future must be equipped with networks designed to be travelled on foot for the daily mobility of citizens and the promotion of healthier, more sustainable, and safer day-to-day and occasional movements. Mobility thus becomes the underlying premise for a revitalisation of territories, an essential condition for relaunching society and the economy, and for recovering the sense of community and belonging to place.

To these solutions for people, we must add those for goods and urban waste. Not only will the cities of the future have to facilitate movement on foot or by bicycle, but also properly reorganise the transportation of goods, adopting innovative projects for the logistics and distribution of goods and the collection of waste. These are principally the objectives supported by the European Commission that, in December 2020, approved the *Sustainable and Smart Mobility Strategy Action Plan*.[5]

The logistics and transport sector is in the throes of a rapid technological and digital transformation. Consumer requests, above all during and after the pandemic, have increased significantly, both in terms of the number of deliveries and the speed at which they must be completed. This imposes the optimisation of storage operations, the management of vehicle fleets, and a reduction in the overall cost of transport—also to compete with well-known big-players like Amazon. While these aspects linked to the transport of urban goods are based principally on the decisions of individual companies (concerning their supply chains and funds for investments in research and innovation). These latter are influenced in turn by legislation and key regulations issued by central and urban governments (Browne and Allen 2011).

Some opportunities are being opened up by technology and Artificial Intelligence. For example, companies like FedEx, Amazon and DHL utilise drones to deliver packages. Projects like HARMONY (pilot project in Trikala, Greece)[6] support the delivery of medicines using drones from the urban centre to pharmacies in surrounding rural areas, completely redefining the role of moving goods in the city. Also, not to be underestimated, is the fact that logistics and goods transportation sectors could be subject to notable innovations thanks to the use of autonomous vehicles. They will help navigate traffic (suggesting the best routes for the most rapid delivery of goods from the warehouse to the client), improving safety for drivers by identifying obstacles, monitoring the state of motors, reducing speed and traffic accidents, and diminishing emissions of CO_2.

These innovations will most likely take form in the decades to come. It will take some time before we feel their effects at the urban, not to mention territorial scale. Despite the drive toward new models of urban mobility being necessary because vehicular transport, other than being damaging and dangerous to human health and safety, is also an evident factor of discomfort in the perception of public urban spaces. For example, the number of parking spaces for deliveries is not sufficient for satisfying the needs of delivery trucks, obliging shippers to double park or park on sidewalks and in bicycle lanes, robbing space from pedestrians and cyclists. The lack of space shifts delivery operations into the street, congesting traffic and generating

---

[5] https://transport.ec.europa.eu/transport-themes/mobility-strategy_en.

[6] https://civitas.eu/news/harmony-projects-sees-first-drone-delivery-for-medicines.

potentially dangerous situations for other users. What is more, the design and position of loading/unloading areas in many cities is often inadequate: stopping spaces too small for trucks and their equipment; or protected based on a fragmented vision, often a response to the demands of a local shop owner and not of a larger scale planning effort.

The impact of the goods transportation and logistics sector in cities is enormous. The CIVITAS[7] consortium (Stefanelli et al. 2020) has recently produced a manual that assembles a series of best practices and solutions which local governments can consult when dealing with this situation. The needs and requisites of logistics differ from city to city in relation to specific local characteristics (scale of the city, dimension and structure of the urban centre, extension and hierarchy of the street network, etc.) require an in-depth study of the most suitable solutions; especially concerning the nature of the principal problem the measure intends to confront (congestion, inadequacy of infrastructures, pollution, noise or safety); at the level of investment required by public authorities (low, moderate or high); and the time for its implementation (short-, medium- or long-term).

4.1.3   Cities Accessible to All

Some urban spaces can feel hostile and excluding. It is not only a question of exclusion caused by economic, religious, or cultural conditions but also by psycho-physical conditions that distinguish between men and women based on their capacity to autonomously access the city's spaces and services. Cities are centres of social interaction where people live and interact with one another. For this reason, they must be accessible to everyone equally. The question is one of inclusion: differences should not be intended as unfavourable factors, but instead as a reason for progressing, improving and discovering new potentialities for the city and its citizens. For cities to be centres of socialisation, where the interaction between individuals is expressed at its highest level, they must be truly accessible and offer everyone the same opportunities to exchange ideas, values, and relations. In an accessible city, everyone (including those with psycho-physical and sensory disabilities) can move comfortably in the streets, with or without public transportation, enter and exit buildings, participate in events and study or work without difficulty.

---

[7] CIVITAS is one of the flagship programs helping the European Commission achieve its ambitious mobility and transport goals, and in turn those in the European Green Deal. Since its launch in 2002, CIVITAS has advanced research and innovation in sustainable urban mobility and enabled local authorities to develop, test and roll out measures via a range of projects. A series of ten thematic areas underpin these (https://civitas.eu/about).

The city is truly accessible when it responds to the needs of its inhabitants through quality design. That is when urbanism aims to reduce existing obstacles and, above all, to augment the possibilities to use the city in the most just and simple manner, with limited effort. This means: (i) limiting elementary obstacles for people (uneven sidewalks, rubbish bins, benches, trees or telephone and/or light poles in the middle of paths); (ii) eliminating physical barriers for persons with impeded or reduced motor capacities; (iii) and add information (visual, tactile, acoustic) for persons with sensory limitations. The objective is to make the urban environment sufficiently communicative and capable of guaranteeing autonomous mobility, orientation and the fruition of spaces in total safety and comfort. In fact, an 'accessible' context has the greatest effect on how individuals are inserted in community relations.

The mobility network is an issue to be approached at various scales to offer solutions for moving through and using the city and the territory suited to every need. Whether for students or workers, tourists or lovers of sport, the elderly or the disabled, urban mobility policies must target two priority goals beyond inter-modality: sustainability and capillarity. Many cities have been working to meet the first objective. For many years, cities substituted old fleets of trains, trams, metros and buses with more ecological and sustainable vehicles. They introduced shared mobility systems (bike-sharing, car-sharing, scooter-sharing) and new devices such as electric scooters, segway, monowheels and hoverboards. Regarding the theme of capillarity, instead, there is still much to be done. The extension of public transport is anything but simple, fast and inexpensive; above all, it may not be feasible in dense and historical areas, where the need to experience the public space of the street is more strategic. In fact, in these urban fabrics, public transport supply must be integrated with a broader availability of pedestrian areas to increase options for travel and accessibility to collective services and facilities. Because moving freely in the city is also a matter of social equity (Secchi 2011, 2013; *Urbanistica* 164).

The network of bicycle paths is an infrastructure (often a protected lane) that, in very busy and densely inhabited areas, must also adapt to sharing space with the street together with other users. Not only the able-bodied, but also the visually impaired or adolescents on skateboards and roller-skates, or in wheelchairs with their caregiver or babysitter and children with strollers and prams should enjoy free movement. In any case, bicycle paths are required increasingly more often to meet 'everyone's needs,' as in other parts of Europe. In Great Britain, for example, a growing number of people with motor disabilities are approaching the use of two/three-wheeled vehicles to rediscover their freedom of movement. Thanks to the awareness-raising campaign of the association *Wheels for Wellbeing* (created in 2007) best practices for 'inclusive cycle mobility' have spread among public governments. In Italy, associations and institutions (e.g., CRIBA—*Centro Regionale d'Informazione delle Barriere Architettoniche*, created in 2000 in the Region of Emilia Romagna) are moving in this direction because when we speak of disability we do not intend only the country's three million disabled (ex-law 104/1992), but the almost ten million people with mobility issues or who suffer from minor disabilities[8] (ISTAT 2019).

---

[8] Referring to the UN Convention on the Rights of Persons with Disabilities from 2006 (ratified by Italy in 2009) "Persons with disabilities include those who have long-term physical, mental,

The 'right to accessibility' is a topical issue for the breadth and complexity of the questions to be faced—disability is not an objective and immutable phenomenon (Merlo 2015)—, and one that in Italy has its first legislative references in Law 13/1989, titled *Disposizioni per favorire il superamento e l'eliminazione delle barriere architettoniche negli edifici privati* (Measures for Favouring Barrier Free Access to Private Buildings).[9] With this law, the legislator issued a specific norm for the elimination of obstacles in the indoor and outdoor spaces of buildings, helping persons with motor difficulties (physical and sensory) to access their spaces and functions. At the urban scale, instead, Law 41/1986 (art. 32)—later integrated by Law 104/1992 (art. 24)—proposed a mandatory urban planning tool for all municipal governments: the Barrier Free Access Plan (*Piano di Eliminazione delle Barriere Architettoniche*, PEBA). The PEBA is a strategic plan aimed at eliminating architectural barriers that still exist inside public buildings and along urban paths. Law 104/1992 established the obligation to integrate the PEBA with the Urban Accessibility Plan (*Piano di Accessibilità Urbana*, PAU),[10] a study of urban spaces focused on the creation of safe paths accessible to everyone. However, it was only with the Decree of the President of the Italian Republic dated 24 July 1996, n. 503, Standards for the Elimination of Architectural Barriers in Public Buildings, Spaces and Services (*Regolamento recante norme per l'eliminazione delle barriere architettoniche negli edifici, spazi e servizi pubblici*), that the theme of urban accessibility was fully introduced in Italian legislation, with particular attention toward pedestrian spaces (art. 4), sidewalks (art. 5), and pedestrian crossings (art. 6).

In Italy, urban planning currently faces a proper shortcoming in dealing with this theme. It must also be said that legislation in this field is also very out of date, having halted at eliminating barriers, without multiplying 'inclusive solutions.' In terms of urban planning, many cities still lack approved PEBA and PAU, limiting the possibility for truly vibrant and dynamic cities. While, in terms of designing for urban accessibility, the theme of visual architectural barriers, despite affecting the majority of the population (Census 2010), seems to be poorly understood and, as a consequence, not dealt with in operative terms. For example, in Italy, the design response to the needs of safe mobility and orientation of people with visual disabilities appears rather imbalanced and not very inclusive, and certainly not responsive to the principles of Universal Design[11] (an American approach, identified by the architect R. Mace in 1985), Design for All or Inclusive Design (more common in

---

intellectual or sensory impairments which in interaction with various barriers may hinder their full and effective participation in society on an equal basis with others".

[9] The Decree issued by the Ministry of Public Works on 14 June 1989, n. 236 titled *Prescrizioni tecniche necessarie a garantire l'accessibilità, l'adattabilità e la visitabilità degli edifici privati e di edilizia residenziale pubblica, ai fini del superamento e dell'eliminazione delle barriere architettoniche* implemented Law 13/1989. Article 3 defines three general criteria for defining a space (built and open) of quality: accessibility, visitability and adaptability. In particular, "accessibility reaches its highest level when it consents the total fruition of one's surroundings".

[10] The PAU is a technical document that illustrates the relations between buildings, infrastructures and users, pointing out risks, obstacles or other sources of embarrassment or discrimination.

[11] https://universaldesign.ie/What-is-Universal-Design/.

Europe). These principles suggest a design of environments usable by everyone, in the broadest sense of the term, without the need to make adaptations or introduce special devices. These principles define different aspects of the quality of our environment, which can be measured/evaluated and thus designed and applied to improve inclusion. As suggested by Mosca and Capolongo these are three aspects of quality: "(i) Physical-Spatial Quality: the capability of the environment to foster easy, comfortable, functional, and safe use of space and objects. This means being able to physically interact with a system; (ii) Sensory-Cognitive Quality: the living environment's capability to foster orientation, comprehension of the service, and comfort of users. Finally, it refers to the features that impact peoples' senses and cognition; (iii) Social Quality: the ability of the environment to enhance and include. All of this considers emotional stimuli and social integration among users" (Mosca and Capolongo 2020).

Additionally, we emphasize that the effort to create cities accessible to all begins with raising the awareness of technicians and citizens about the construction of public space without 'minor or temporary barriers,' such as job sites, waste or urban furnishings that limit the full enjoyment of the city by everyone. Of notable interest in relation to this latter aspect, is the response offered by Paris with its Accessibility Plan for Roads and Public Spaces (*Plan de mise en accessibilité de la voirie et des espaces publics, or Pave*), which projects the city into the year 2024 with new accessibility for all (Rossi 2022).

4.1.4 The Street as Community (Public) Space

Public space has shown itself to be the principal ally of cities in the fight against the pandemic. At the neighbourhood level, public squares, parks, parking and the street have become the space for emergency services (temporary hospitals, vaccine centres, storage and structures for distributing food) and supporting social resilience and people's wellbeing.

Open spaces (plazas, streets, parks) have generally played an emblematic role in reducing the negative impacts of the pandemic on society. For example, during lockdowns, Denmark permitted physical activity and socialisation at a distance by keeping its public spaces open and secure. They did so thanks to a few clear rules on physical (and not social) distancing, contributing to the community's cohesion, alleviating stress and playing an essential role in childhood development. Additionally, they noted that public spaces were used equally by men and women unlike in the past, thus reducing gender disparity (Gehl Architects 2020).

Despite this, the health crisis underlined diverse shortcomings in the public spaces offered by existing cities, which must be dealt with in the short-, medium- and long-term, including accessibility, flexibility, design, management and maintenance, connectivity, and equitable distribution in the city (UN-Habitat 2020).

Collective life is manifest in the myriad variations of public space through programmed or unplanned contacts, and this attributes a sense of character to the city

(Marchigiani et al. 2017). Public space is where people are stimulated by the environment (also social) and where, more than elsewhere, we find the surprise effect: the wonder of the unexpected. Serendipity,[12] often lacking during the periods of lockdown and periods of restrictions imposed to limit contagion, disappeared from people's lives and, consequently, from the spaces of collective life, transforming public space into areas that offered no possibility to build ties. As Aristotle wrote: "man is a social animal" in need of continuous relations with other individuals for self-affirmation and evolution. In this relationship, man requires physical action because he is "not in space, but inhabits it," models it, adapts it, and shapes it to stimulate moments that enrich his intellect and spirit. An example is offered by J. Gehl's studies of public space and the difficulties in allowing people to stop in an area (public) rather than simply passing through it. Stopping is what nurtures control, safety, occasions for encounters and the creation of relations. Therefore, it is no accident, that many streets are happily transformed from places of passage into places where people stop and gather.

The street is considered a public space *par excellence* (that is, perceived as the 'space of everyone'), in which to exercise direct forms of exchange and informal and convivial relations, and to hold civil demonstrations. Post-pandemic normality has led to a re-evaluation of the street as a space of walking (for psychophysical wellbeing), observing (here wind the return of serendipity), and re-establishing a dialogue with/in the community (for social wellbeing).

During the lockdowns—which emptied the streets—vehicular traffic and public transportation were drastically reduced, restoring the focus on strategies for recapturing space to offer people more areas for socialisation and safe movement, as well as facilitating travel by bicycle and on foot. The public space of the street, particularly its temporary use (the ability to change and adapt over time in response to different needs), became an essential element for increasing the urban quality of cities. This element must once again be the strong focus of a program of interventions and a project for public space (Indini 2015), also beginning with the reconsideration of its material surface (Dell'Aira 2017). Reasonable assistance was offered by various manuals worldwide (from New York to Boston, from Berlin to London, from Moscow to Shanghai). Here, we would like to mention the latest document by the City of Milan, in collaboration with AMAT, from November 2021 and titled *Spazio pubblico. Linee guida di progettazione*, offering a broad range of solutions for the public space of the street.[13]

Completing the network of ecological public transport and bicycle itineraries with extended and interconnected areas for pedestrians and social interaction is a priority for governing the territory. The goal is to offer alternative means and paths that permit urban systems, inherently complex, to deal with possible shocks. Primarily to remain an efficient system, even during interruptions or limitations caused by

---

[12] The term was coined by Horace Walpole during the eighteenth century.

[13] https://issuu.com/saf_arch/docs/spazio_pubblico_-_linee_guida_di_progettazione.

different factors, either external (flooding or blackouts) or internal (ruptures or maintenance/improvement works). Urban design that offers multifunctional spaces for mobility participates in realising more resilient cities.

The relationship between urban form (a determinant factor for the resilience of cities) and mobility has deep roots in history. It is characterised by reciprocal influence: streets that design new expansions and patterns of settlement that determine a precise hierarchy of roads (Balletto 2022; Ravagnan et al. 2022). This relationship has also created polarities and centralities, distinguishing the use of the street and the functions lining it in homogeneously, giving priority to pedestrians in central areas. There is more. In many cases, the 'city effect' is determined precisely by the convergence of streets toward specific places, with a radial system of paths defining urban centres. Today, the roads in many European cities continue to follow a radial structure. With it, most public transport systems principally link urban centres and peripheral districts.

In Italy, urban centres were the first places to be pedestrianised after the Second World War, and in particular after the 1960s. At a time when historic centres were assuming their identity e autonomy—in terms of conservation, protection and planning—public squares and some heavily congested streets were closed, from Siena to Bologna, from Milano to Novara, to the emblematic experience of the pedestrianisation of Via dei Fori Imperiali in Rome at the end of the 1980s, or Piazza del Plebiscito in Naples in 1994 (Donati 2017). In Europe, a number of experiences had already anticipated these events, in Rotterdam, Copenhagen, and in other megalopolises on the other side of the ocean. However, closures of roads to vehicular traffic in New York (2009), Vancouver (2010), Mexico City (2010) or Los Angeles (2013) have taken place only in the past 15 years (Spirito 2013; Raganà 2016). Today, this practice is accompanied by others intent on increasing portions of the streets where pedestrians are the priority.

Different strategies for requalifying the spaces of the street shared by different users have been sedimenting for years. Here we find interest on the part of the public sector in developing a collective space for various activities, and not solely for mobility. Shared street, *woonerf*, *begegnunzone*, *verkehrsberuhigter bereich*, *zone de rencontre*, Living street and *zone a traffico limitato* are all operations that since the 1970s have sought to support a promiscuous use of the street by pedestrians, cyclists and motorised vehicles.

The typology of intervention known as the 'shared street,' utilised for the first time in the 1980s in Chambery, France, transforms the street into a public space 'at the same level,' establishing a coexistence between pedestrians, cyclists and motorists. The elimination of sidewalks, curbs, divisions between users and lanes creates a single space regulated by a fundamental premise: attention and priority are given to the weakest. The Peltzman effect (1975) is at the base of these interventions: people tend to regulate their behaviour in relation to a perceived level of risk; in this manner, both motorists and pedestrians behave with greater caution when they feel vulnerable, as in the case of space without barriers or separations.

Another concept that works with the promiscuity of street users is known as *woonerf* (Dutch for 'shared area'). *Woonerf* is a street where precedence is assigned to pedestrians and cyclists and where, thanks to a series of devices, motorists are forced to adopt a more prudent style of driving. This intervention was initially conceived in the Netherlands in the 1970s/80 s, and has since spread across Northern Europe and North America. Developing in different countries, *woonerf* has taken on slightly different connotations and regulations: in Switzerland it is known as *begegnunzone* (zone of encounter), in Germany as *verkehrsberuhigter bereich* (moderated traffic area), while in France the term *zone de rencontre* (areas of contact) is utilised, and Anglo-Saxon countries refer to living streets. Italian legislation does not contemplate the creation of *woonerf*, though they can be considered similar to the country's 'pedestrian zones,' or '30 km/h zones' and 'limited traffic zones.' Despite the diverse linguistic variations, the essential point is an inversion in the hierarchy among subjects, without entirely banning access to motorised vehicles. Pedestrians are always given precedence wherever motorised vehicles are present, and the latter must proceed at walking speed (Guzzabocca and Legoratti 2021).

Even the most recent experiences in 'tactical urbanism' have continued to work with the re-appropriation of public space by pedestrians. This approach is based on the idea that a street or square arrangement can be changed rapidly and at a low cost. This approach includes two main principles: the 'waterfall effect' of urban planning actions (interventions must manage to trigger successive processes of requalification); and the temporariness/reversibility of interventions that need to be realised quickly and used as design-tests for new urban layouts. Actions that, eventually, maybe consolidated over time.

However, these operations, which differ in time and space, confirm the numerous useful aspects derived from the practice of walking in the city and consequential benefits: social (safety, placemaking, social cohesion/equality, health, and wellbeing), economic (local economy, urban regeneration, city attractiveness, cost-savings), environmental (ecosystem services, virtuous cycles, liveability, transport efficiency), and political (leadership, urban governance, sustainable developments, planning opportunity). In short, "Mobility is intrinsic to the quality of life experienced in cities. We now have the opportunity for human-centred design, to place people back at the heart of our cities. A walkable city is a better city and putting walking first will keep our Cities Alive" (ARUP 2016).

The street can thus be considered an 'agent of proximity' that operates in a stable manner for the communities and citizens of the quarter, bringing social capital to the surface, creating occasions for building relations, trust, and processes that enable bottom-up innovation and the capacitation of citizenship. Therefore, the governance of the territory for the city of the future must respond to the request for the 'right to space' that activates forms of bottom-up social protection and augments the quality of dwelling.

If on the one hand, in the contemporary city the quarter is not a community in which we can find the conditions for sharing values, norms and experiences, beyond the reciprocity of practices (Borlini 2010); on the other hand, the 'design of the soil' (*progetto di suolo*) (Secchi 1986, 2006)—that involves *in primis* the 'depth of the street' (Secchi 1989)—can limit the differences among social groups living here. Groups that are also recognisable in prevailing architectural and urban models, as well as the cost and prestige of homes. For collective use like the street, the design of the ground level can increase the degree of material and immaterial connections with the rest of the city and, therefore, with the community (Babic 2021). Furthermore, the street has already been the subject of numerous studies that demonstrate its central role, at various scales, in contrasting the effects of climate change, for example, heat islands and flooding (Ali-Toudert 2007; Tamminga et al. 2020). Properly designed—in particular with soft mobility in mind—streets can help build the resilience of cities (Sharifi 2019; Abastante et al. 2020). By increasing comfort (climatic, environmental, spatial, functional, performance-based), the design of the street can return to once again being a project for public space suited to hosting individual activities (walking) and collective activities (socialising), restoring a sense of conviviality to neighbourhood practices.

4.2 An Ecological (Urban) Transition with Green and Blue Networks

4.2.1 Toward New Urban Models (Healthy, Sustainable, Intelligent, Adaptive)

A recent publication by the European Environmental Agency (EEA 2021) states: "Cities are facing a triple crisis in the wake of the pandemic: tackling the health impacts of Covid-19; dealing with the climate and ecological emergency; and addressing social and economic inequality. Despite these challenges, cities have the potential to become a major driving force for a green and just recovery in Europe— provided that they are actively involved in the decision-making process from the beginning. Although it is too early to know what the longer-term legacy of the pandemic will be for urban environmental sustainability, it is clear the unprecedented EUR 1.8 trillion stimulus package agreed by the EU will reshape cities in fundamental ways. Infrastructure investment will play an important role in stimulating urban economic activity after the crisis, creating an opportunity to align the recovery with climate, environmental and social equity agendas in cities. This will need to be accompanied by better integration of policy sectors and actions to maximise co-benefits. Key opportunities for a green and just recovery are found in the following sectors: rethinking urban mobility and land use; retrofitting the urban building stock; enhancing the role of green infrastructure and nature-based solutions; and transforming urban food systems and the circular economy."

Urban systems can therefore be considered spatial entities in which to invest economic and social resources to subvert the fate of the Planet. For the cities of the future, places of connectivity, creativity and innovation, and services, the objectives of efficiency and competitiveness are both essential for achieving a new urban model. To be competitive, cities must function efficiently and make the most intelligent use of available public and private resources. They must know how to increase their resilience (to challenges to health, the environment, climate, economic, social, digital, and technological) and concretely implement sustainable urban regeneration projects. In fact, by promoting innovation, we can support the transition (not only ecological) of the cities of the future, according to the principles of the European Urban Agenda 2030 and reiterated in the Leipzig Charter 2020,[14] which promotes cities that are more ecological, inclusive and cohesive, productive and connected. Cities are, therefore, the target for innovations in products (quality and living conditions) and processes (urban metabolism, development of settlement, governance among actors), that is, centres toward which to converge the substantial innovations of the future (cities as a true model of healthy, intelligent and sustainable settlement). A new urban model that absorbs the concepts 'Smart City,' 'Safe City' (Ristvej et al. 2020), 'Adaptive City' (Manigrasso 2019), and 'Smart-Eco-City' (Finka et al. 2016). A similar model requires an urban strategy based on green and blue network planning, providing for the expansion, protection and connection between green areas, water bodies and the reuse of waste landscapes (drosscapes). The urban model also requires soft mobility networks, sustainable and less energy-intensive mass transport and the mixed-use of land innervated by tangible and intangible services (Gasparrini 2015; Gasparrini and Terracciano 2017; Andreucci 2017; Ricci 2021). At the same time, it requires a precise design of space at the 'human scale' by combining elements of technological innovation (green solutions such as passive houses or green roofs/walls) and stimulating behavioural changes in people (sustainable travel, waste separation and lower energy consumption) (Bibri and Krogstie 2020).

On 13 July 2021, Italy's National Recovery and Resilience Plan (*Piano Nazionale ripresa e Resilienza*, PNRR)[15] was definitively approved by the Council of Ministers, incorporating the proposal from the European Commission to provide access to the EU's Next Generation funds. The Plan develops in six Missions and pursues three principal objectives. With a near-term temporal horizon, the first is the remediation of the economic and social damages caused by the pandemic. In the medium-long

---

[14] https://ec.europa.eu/regional_policy/sources/docgener/brochure/new_leipzig_charter/new_lei_pzig_charter_en.pdf.

[15] https://www.governo.it/sites/governo.it/files/PNRR.pdf.

term, the Plan confronts territorial gaps, gender disparities, weak growth and productivity and low human and physical capital investments. Finally, the resources made available by the Plan will bring an impulse to an ecological transition aligned with the need for cities that are more resilient, sustainable, healthy, intelligent and socially inclusive.

The Plan forecasts more than 190 billion euros of investments (plus a further 30 billion approved by the Government in a complementary fund). In particular, Mission 2, 'Green Revolution and Ecological Transition' will receive almost 70 billion euros for actions focused on a circular economy and sustainable agriculture, renewable energy, hydrogen, sustainable mobility and networks, energy efficiency and building refurbishment, the conservation of the territory and water. Mission 3 'Infrastructures for Sustainable mobility' will receive some 32 billion euros for themes like highspeed rail and road maintenance, and intermodality and integrated logistics. The quantity of investments available at the European level represents an incredible opportunity, though it needs to be coordinated to be feasible at the territorial level. This is why, in December 2021, the Italian National Planning Institute (*Istituto Nazionale di Urbanistica*, INU) unanimously approved and submitted to the Government, in the form of a legislative proposal, the Integrated Regional Program (*Programma Integrato d'Area*, PIA).[16] The PIA is a tool for promoting the resources of the PNRR at the territorial level, effectively guiding strategies of urban regeneration and ecological transition; and, at the same time, mobilising private funding.

4.2.2   Green and Blue Infrastructures

International orientations consider green (and blue) infrastructures an essential and strategic investment in the ecological transition of the city (the green shift), even after the pandemic. This concept was developed in the USA[17] beginning in the 1990s, while Europe officially introduced the term in 2009 in a guide to climate change (CEC 2009), and in 2013 the European Commission published one of the first documents on this issue (EC 2013). The term 'green (and blue) infrastructure' derives from a progressive inclusion of ecology within urban planning. In the history of modern urbanism, other terms refer to the topic of environmental connections and the fruition of the landscape in the city. Passing from hygienic and aesthetic dimensions to another dimension, more centred on ecological and quantitative value (Campos Venuti 1983,

---

[16] https://aliautonomie.it/wp-content/uploads/2021/12/INU_Piano-Integrato-di-Area-10.12.2021.pdf.

[17] http://water.epa.gov/infrastructure/greeninfrastructure/.

2004, 2009; Oliva 1993), and later on the ecosystemic and qualitative value of the environment. They include the concepts of the 'green belt,' 'greenway' and more recently 'ecological networks/ecological corridors.' The term Greenway is derived from the fusion between 'green belt' formulated by E. Howard (to protect residential settlements against industrial expansions, uniting the commodity of urban streets with the pleasures of the countryside), and 'parkway' formulated by F. L. Olmsted and C. Vaux, while designing Central Park in New York (to indicate the entrances to urban parks with trees and vegetation) (Panzini 1993).

Green infrastructures are multifunctional and interconnected systems of natural and semi-natural areas (agriculture, parks, gardens), vegetation (rows of trees, forests, hedges) and green constructions (green roofs, preamble paving, rain gardens, green trenches) that bring environmental, climatic, economic and social benefits to urban systems (John et al. 2019). These benefits include processes of infiltration, sedimentation, evapotranspiration and/or recycling of rainwater; reductions in airborne pollution and dampening of noise; mitigation of heat island effects that contribute to a reduction in energy demand. Blue infrastructures (blue spaces) are those most directly linked to water, such as rivers, canals, lakes, and the sea. They represent a structural and functional network of equal importance to the natural and anthropic environment. In fact, according to the biophilia hypothesis, developed by Edward O. Wilson in the 1980s, during their evolution, humans being have matured a strong connection with nature, moving the subconscious toward the search for natural environments, including green and blue spaces, to rebalance their psycho-physical wellbeing (Wilson 1984).

Like ecological networks (that is, the planning/design of environmental connections, opportunely equipped to permit exchanges between local fauna and favouring the conservation and enrichment of genetic diversity in the territory), Green and Blue Infrastructures (GBI) adhere to a multi-scalar vision of design. These infrastructures play a fundamental role in urban systems at the territorial scale, re-connecting large extra-urban green areas, while encouraging sustainable and informed use. At the same time, GBI are also strategic at the local scale, where ecosystemic relations and social, health, and economic benefits are even more urgent.

Today the challenge is to utilise the concept of GBI at both the strategic level (promoting policies that target the sustainability of settlement systems) and at the spatial-design level by developing and implementing the concept of 'networks.' In fact, to obtain the benefits of nature in the city, the presence of parks and gardens is not enough; they must be physically connected with one another and with other green and blue systems (Chiesura 2018). Connections in a dense and stratified urban fabric must consider all linear elements with the potential to join the principal nodes in the network. These elements include streets, that is, the public space in which it is possible to add more natural solutions (trees, rain gardens, green parterres, etc.).

They contribute to ecological continuity and to the fruition of the city using means with a low impact on energy and the environment. Ecological and environmental continuity must move hand-in-hand with the promotion of forms of mobility that offer alternatives to private vehicles, as the redesign of the streetscape will be ineffective if circulation remains dedicated prevalently to automobiles.

GBI can and must be imagined at different scales: the building, the neighbourhood, and the city. In particular, the design of these infrastructures must feature a series of interconnected features adapted to the context into which they are inserted. Some solutions, for example, green roofs, water squares, or temporarily floodable areas, filtering systems such as gardens, parks, rows of trees, etc., are impossible to realise separate from the territory. It is enough to think of the delicate nature of historic centres and ancient nuclei. Similarly, operations to de-seal paved soils—to create space for green systems and surface water creating new urban landscapes—represent a difficult challenge in the existing city's dense and compact fabrics. Hence, the integrated design of green and blue networks, focused on the regeneration of the existing city, must be pursued to bring about a true green shift and offer public spaces accessible for social life, recreational activities, wellbeing, and human health and safety, focusing a great deal of attention on the context into which projects are inserted (ARUP 2014).

The benefits produced by the forecasting and realisation of GBI affect both the prevalently ecological-environmental sphere and social issues and healthcare (Ramboll 2016; Brown and Mijic 2019; WHO 2021); because the innovation to the concept of GBI includes the offering of Ecosystemic Services provided by the soil (Ronchi and Arcidiacono 2021). GBI are key to urban policies, urban planning and urban design, and essential for conserving Natural Capital and improving people's quality of life. This approach should be incorporated within territorial programming devices and land regulations, which rarely consider the territory's ability to produce multiple benefits (Arcidiacono and Ronchi 2021).

4.3   Active Mobility for Psychophysical Wellbeing and Sustainability

4.3.1   Active City and Salutogenic City

The environmental dimension of cities is at the heart of urban regeneration strategies to reduce both the negative externalities of traffic and global warming, improve safety for city users, bring a new impulse to physical activity and improve people's health. In particular, the issue of 'active mobility' is essential for supporting a real change in behaviour and improving the quality of cities and peoples' psycho-physical wellbeing (Fior et al. 2022b).

The Covid-19 pandemic obliged children, adolescents, the disabled, and the elderly, but also students and travellers, to adopt a sedentary lifestyle (Zheng et al.

2020). Returning to daily sports activities may not be all that simple. While shopping for food using e-bikes, visiting a community centre after strolling beneath a row of trees, or sharing a daily path (home-work or home-school) by bike or on foot with friends, can be a way of facilitating a healthy and sustainable lifestyle. However, it necessitates adequate spaces and infrastructures be put into practice.

The environment in which people live, act and move, has direct impacts (air quality, climate, noise, traffic, natural risks) and indirect impacts (conditions of dwelling, socialisation, accessibility, fruition) on human health (Barton and Grant 2006; Capolongo et al. 2020). What is more, related literature confirms that the consistency and density of the landscaping and the design of public space (materials, urban furniture, dimension, services, and facilities) intensify the use of the city by pedestrians and cyclists (Vich et al. 2019). Therefore, increasing the networking and continuity of these spaces helps improve the health of citizens because it encourages their freedom of movement.

A 2016 study by Sallis and colleagues demonstrated a significant positive and linear correlation between four environmental characteristics and physical activity. These environmental characteristics are residential density, the density of public transportation stops, the density of intersections, and the number of green spaces (Sallis et al. 2016). The study reveals how it is not enough to limit risks to health in the contemporary city, but instead that we must promote health in cities. The relatively recent concept of 'Healthy, Active City' is tied to a growing interest in studies and research into the cause-effect relationship between the city and health. In particular, the concept shifts from 'caring pathologies' (a prevalently medical approach) to 'prevention' pursued through healthcare policies that include actions in urban planning (Dorato 2020). The passage from a healthy city to a "Salutogenic City" (Antonovsky 1979) implies working on a behavioural, and thus cultural change that induces people to become those most responsible for improving their living conditions. From the perspective of prevention, this applies to anticipating the onset or increase in chronic-degenerative pathologies (typical of contemporary society).

The World Health Organization (WHO) has noted that the global cost of physical inactivity is estimated to be INT$ 54 billion per year in direct health care caused by excessive sedentariness (WHO 2018). Diseases (such as obesity, diabetes, cardiovascular illnesses, depression, etc.) could be easily reduced if the population regularly practiced physical activity. This latter term refers to any bodily movement produced by skeletal muscles and requiring the use of energy. Research financed by the International Sport and Culture Association demonstrates that 500,000 Europeans die each year due to a lack of physical activity, as 25% of the European population does not engage in sufficient movement. However, the most disturbing aspect is that 80% of adolescents are inactive (ISCA 2015). Educators and decision-makers play a determinant role in supporting a radical lifestyle change and, mainly, in promoting healthy and active lifestyles to reduce public healthcare costs. According to indications provided by the World Health Organization, 2.5 h a week of activity of moderate-intensity, or 20 min a day, are sufficient (WHO 2009, 2010). The Italian National Institute of Health has declared "moderate physical activity to be physical activity that comports a slight increase in breathing and the heartbeat and light perspiration for its quantity, duration, and intensity. Examples include walking at a brisk pace, riding a bicycle, practicing light gymnastics, dancing, gardening or doing housework, such as washing windows or floors".[18] Riding a bicycle or walking short distances can be considered moderate physical activity. In terms of urban planning, to reduce the mortality and morbidity caused by excessive sedentariness it would be sufficient to support actions that increase active mobility and walkability.

The term 'active mobility' refers to a practice that promotes the use of the bicycle and movement on foot. Promoting active mobility—cycling or walking—for at least 10 min at a time and a total of at least 150 min per week guarantees the achievement of minimum levels of physical activity recommended by the WHO, independent of physical activity practiced during free time or at work. Compared to sport, active mobility is suited to a broader population range. It requires less time and motivation and can be easily integrated into peoples' daily activities—traveling to and from work, grocery shopping, accessing services, and accompanying children to school—but also activities during their free time, such as strolling and cultural excursions.

For the world of design, it is worth emphasising that the WHO's *Global Action Plan* to promote physical activity (WHO 2018) reveals that simply building a bicycle path is not enough to make people use it. Other determinant factors include the density of buildings and functions (rich and variegated urban environments are chosen more for the possibilities of movement), the presence of curated landscaped spaces (that, for example, attract children looking for a place to play, the elderly seeking a place to relax and adults wishing to engage in sports), the presence of intermodal mobility offered by public transport (which facilitates the fruition of the expanded contemporary city); as well as the perception of safety, above all for the elderly, offered by well-lit and clearly indicated open space, without, for example, urban voids, degraded or under-utilised areas.

4.3.2   Walkability

The walkability of urban space is a multi-dimensional concept that can be difficult to measure; it can be linked essentially to places in the city whose performance characteristics (in terms of accessibility, functionality, and density) invite people to walk (Dovey and Pafka 2020). An interesting application developed by the Institute for Transportation & Development Policy[19] compares the walkability of the world's leading cities to disseminate the principle that pedestrians should be a priority in urban policies and projects (Pedestrian First). The study shows how Europe's principal urban areas offer different levels of accessibility to schools and healthcare. In Paris, 85% of residents live within less than a 1 km walk from these services, in Lisbon this value is 84%, in Brussels 83%, in Milan and London 80%, in Oslo 78%, Berlin 77%, in Barcelona 72%, Prague 71%, Helsinki and Valencia 60%, Zagreb 59%. In

---

[18] https://www.epicentro.iss.it/passi/indicatori/attivit%C3%A0Fisica.

[19] https://pedestriansfirst.itdp.org/city.

comparison, in Amsterdam, this number drops to only 52% of residents, while it reaches 41% in Naples.

Promoting active mobility and walkability offers people a valid alternative to public transport and vehicular movement and a real possibility to practice physical activity. The result is more liveable and safer cities. Since the spread of the Covid-19 pandemic, the structures of mobility systems, above all in large cities, have changed drastically. Public transport, forced to reduce its offering by 50%, cannot properly satisfy demand, while private automobile traffic and dangerous risks to the environment and human health have increased. The best alternative is represented by bicycle-pedestrian mobility. However, it is often hindered by the low capillarity of networks.

Nonetheless, reflections on this issue demonstrate that what makes a city attractive to walking is not only the availability of infrastructures (sidewalks, pedestrian zones, public squares). Conditions of context, such as the number of destinations of interest, the quantity of urban opportunities offered and the distance between them, the quality of routes (variety of functions, perceived safety, environmental comfort) are also fundamental. The correct quantitative–qualitative evaluation of the incidence of these factors and their combination in determining urban quality is essential to scientific research. Despite this, in real life, what counts is the twofold role offered by areas for pedestrian use as an increase in individual wellbeing (walking reduces the onset of many chronic-degenerative pathologies) and as an increase in environmental wellbeing (social control, reduction in temperatures, ventilation, etc.) (Blečić et al. 2015).

Certainly, spatial disciplines such as urbanism are interested in how to concretely approach the improvement of the quality of life through the promotion of spaces suited to pedestrian mobility. Cerasoli proposes returning to Marcello Vittorini's theory from the late '80 s concerning the design of a "theoretical grid." It is the schematisation of the elementary urban unit (the quarter), in which to re-establish hierarchies and centralities and the quality of the urban environment. Among these centralities, Cerasoli underlines public space of relation (streets, public squares, tree-lined avenues, and covered galleries) as a device that regulates the recognisability and qualification of surrounding fabrics (Cerasoli 2015).

4.3.3 Proximity Healthcare

The pandemic has once again vigorously exposed the need to create sustainable mobility networks, now a priority in many cities. There is more. The closure of cities also revealed the need to evaluate the performance of healthcare systems and assisted care facilities (for example nursing homes), in terms of accessibility, quality of care, and efficiency. In particular, the pandemic has shed light on the need for a reorganisation, at the territorial level, of assisted care for managing chronic pathologies and targeting 'proactive patient management' (Chronic Care Model)[20] working principally in two directions. On the one hand, expanding the preparation of programmes dedicated to chronic illnesses to other territorial structures (social and volunteer), leaving hospitals to manage more acute cases and the local territorial network (the *case della salute*, outpatient centres) to manage other patients. On the other hand, by making patients aware of illnesses and sharing the responsibility for their health with them. Most importantly, this second point requires an effort toward adaption on the part of cities. This means an improvement in their performance (environmental, infrastructural and settlement-related) translates into the possibility to help people move toward a real behavioural change that targets patient autonomy in synergy with a reconsideration of the offering of healthcare. The passage from *medicina d'attesa* (waiting medicine, when patients wait for medical treatment) to *sanità d'iniziativa* (initiative healthcare, when patients are proactively involved in improving their health) must be accompanied by improvements in urban infrastructures. As urban services that facilitate the ability of patients to autonomously carry out simple day-to-day activities (walking and grocery shopping), they have a substantial impact on reducing lethal chronic illnesses.

In this sense, promoting the regeneration of public spaces, equipping them for walkability and ludic-recreational use, is a vital urban planning action. Particular guidelines certainly include those linked to forecasts for dense and mixed-use blocks (residential/services), though of limited dimensions. The work in Paris with 15-min cities, or the 20-min cities in Singapore and Portland, move in this direction, with the precise intent to contain long-range movements and stimulate pedestrian movement. Additionally, the multi-functionality of public space, which creates opportunities for encounters, leisure and fruition, integrates the principle that attracts people to use the city's public spaces, brings them to life and supports residents' needs tied to the use of essential services and ordinary exercise.

Therefore, the city of proximity is a city that positively influences the rhythm of life in the urban system, reduces speed and increases interpersonal contact. It is a city that also has the potential to become a city that cures, an ecosystem of people, organisations, places, products and services capable of caring for one another, breaking free of a vision according to which assistance is provided uniquely by specialised operators. The term care refers to numerous and diversified activities that vary based on the time required, the level of specialisation of those responsible and the level of responsibility needed: healthcare workers (specialised and professionals), third sector and charity organisations (specialised and non, professionals and non), traditional care communities (families, neighbourhood communities). To regenerate a city capable of caring, we must develop new communities. To create new communities, we require a new generation of collaborative services, distributed across the territory in a proximity system, as "there is no care without contact" (Manzini 2021: 66).

The epidemiologist Marco Geddes da Filicaia (2020) states, "the strength of the Italian National Healthcare System is certainly its accessibility". Indeed, the Italian system differs from other international situations for its being a public service

---

[20] https://www.bmv.bz.it/it/glossario-parole-chiave-e-riviste/chronic-care-model/.

that guarantees all citizens equal access to healthcare. In practice, however, there is a chronic territorial disparity (above all between North and South) in access to services, aggravated by the incapacity to respond to changing demographic, social and environmental conditions. The negative aspects of a hospital-centric system were confirmed during the pandemic, and further aggravated by a situation in which hospitals remained the only point of reference. Furthermore, it also became an effective vehicle for spreading contagion between patients and staff. A network of proximity facilities could have reduced the pressure on hospitals. Many patients who visited emergency rooms—placed in code white or green—could have been cured at home or in other structures without further burdening an already saturated hospital system.

The city, like healthcare systems, must also be able to provide suitable responses to new trends, such as an aging population, a rise in chronic degenerative illnesses and the adoption of sedentary lifestyles. However, it is necessary to rethink urban structures and healthcare systems, building systems capable of adaptation, new programmes and plans, capable of promoting health and suitable lifestyles. Indeed, speaking of 'proximity medicine' means recognising that healthcare has changed, and with it, the means of accessing and using its services and structures.

A recent national study by Scale reveals that the most requested healthcare services are lab tests (66%) and family physician and paediatric visits (64%) offered by public structures or others affiliated with the National Healthcare System (Deloitte 2020). What also emerges from this study is that access to emergency rooms accounts for only 21% of the total demand for healthcare services. While the healthcare system is generally recognised as being among the best in the world, it nonetheless emerges that many patients abandon medical care for economic reasons (more than 25% of those surveyed).

A successive study by Nomisma with Rekeep revealed that avoiding excessive overloading of the central healthcare system (also in the wake of the pandemic) will require improvements to proximity healthcare, in line with the Italian PNRR 2021 that suggests the same. (Nomisma 2021). The study begins with the situation of the healthcare system operating across Italy and the model of healthcare outlined in the PNRR. The focus is on healthcare facilities and defining interventions required by *Ospedali di Comunità*,[21] *Case di Comunità*,[22] and Healthcare Residences (*Residenze Sanitarie Assistenziali*, RSA) to reinforce the territorial healthcare network, together with the relative economic, social and environmental benefits. The healthcare system must readapt to the territory and the concept of health. Achieving this means imagining an integrated network of territorial structures to create a diverse dimension of spaces offering care (Capuano 2020).

---

[21] The *Ospedali di Comunità* are healthcare facilities for medically stable patients who no longer require hospital care but remain too unstable to be treated as outpatients or in-home patients.

[22] The *Casa di Comunità* is a multipurpose facility that offers healthcare and social services under one roof, thanks to the cooperation among and integration of multiple healthcare professionals and outpatient structures. Additionally, the presence of social workers in the *Case di Comunità* makes it possible to reinforce the role of territorial social services (Bartoloni 2021). The social dimension is the distinguishing feature of the *Case della Salute*.

One of the fundamental objectives of healthcare is to bring care as close as possible to patients and not vice versa. As part of a similar operation, the city must guarantee an offering of infrastructures and supporting spaces for territorial care that allows people to access facilities of proximity in total autonomy. Cities, and in particular outdoor and indoor spaces, must become true 'enabling platforms' of this process (Miano 2020) offering solutions that satisfy accessibility to healthcare services and stimulating the autonomous movement of all individuals (the elderly, mothers, the disabled), without forgetting that a healthier city must be backed by a new political agenda that supports feasible urban projects (Toppetti and Ferretti 2020).

4.4 The M4 Stations, Nodes of Slow Mobility

In Milan, the Green and Blue Backbone project of the M4 metro line demonstrates how it is possible to integrate mass mobility, slow mobility and nature-based solutions for regenerating local and metropolitan connections and promoting new, safer and healthier lifestyles. In particular, the design of networks that expand bicycle paths and extend pedestrian areas weave together a system of public spaces closely related to the metro route. This double level of flows (high-speed underground flows of the metro and slow movements of pedestrians and bicycles on the surface) is, above all, the first concrete action for improving the city's resilience. An expansion in the offering of mobility makes it possible to design infinite hypotheses for crossing the city from East to West, reducing possible inconveniences where there is a disservice in public transport for various reasons.

Another critical aspect exposed by the project developed for Milan is the possibility to efficiently integrate actions to adapt dense and historical urban fabrics to climate and environmental changes, respecting their meaning and historical-cultural value. The bicycle and pedestrian paths framework was opportunely verified during site visits and urban-architectural surveys. The continuous and capillary network of paths assumes different thicknesses and consistencies based on the built environment crossed. Thanks to their important width, urban boulevards located in proximity to ring roads or in the city's more modern neighbourhoods lend themselves to the planting of new rows of trees and green parterres as well as gardens and children's playgrounds. Vice versa, historical road networks and those close to the city centre present a narrow section and adherence to ancient buildings. These streets were selected for repaving operations using better draining materials or with a greater albedo to reduce urban heat islands and improve comfort in public squares and pedestrianised areas.

Additionally, the in-depth study of several pilot stations also verified the feasibility of redesigning public spaces, bicycle and pedestrian paths, and the re-opening the city's ancient navigable canals (Cerchia dei Navigli). Green and blue networks were thus successfully integrated within a design that promotes identity of place by exploiting the arrival of the new metro stations. The Vetra station, for example, is situated in a zone of particular historical and symbolic value for the city—home to important basilicas and one of Milan's most beautiful and scenographic park (Parco delle Basiliche)—was functionally and aesthetically integrated within the landscape. In this design experiment, the network of territorial connections was translated into the spatial organisation of a pedestrian hub. In the area near the metro exits, the concomitant forecast to reopen the ancient canals and the lack of an environmental continuity within the park, guided the requalification of public space by suggesting an integration of pedestrian paths and ecological corridors in a much wider bridge. It is almost a suspended public square, with a gentle slope to facilitate its use by everyone. The proposed architectural structure uses different materials (local stone, small stones, glass, and wood) to invite users to cross this space and to stop and enjoy new urban perspectives offered by the reduction in vehicular traffic.

In the Sforza-Policlinico station, instead, the need to make up for the absence of a subterranean connection between the new blue line subway station and the existing yellow line station was the impetus for studying material solutions for surfaces and furnishings designed to help orient (Lynch 1960). These solutions can be reinforced and improved, but they start to sediment a new way of creating public space with an eye on Design for All. The realisation of view cones marked by newly planted trees, the arrangement of bollards for automobiles in the refurbished plaza indicating 'disadvantageous paths for pedestrians,' and the positioning of accurate signage and indications led to the development of a proper operation of Wayfinding Design. The latter is a practice that utilises signs and symbols in a visual and informative manner to assist people when crossing or using a space, or a series of spaces, by creating a true and proper map for the user. Clarifying the physical form (spatial/functional) of place and applying texts to some of its parts, the 'urban rooms' of the Sforza-Policlinico station communicates with users of the subway and the city looking to either reach the station itself, or visit the city's historical and cultural heritage (twentieth century Milanese architecture, churches, museums, hospitals and universities) (Figs. 4.1, 4.2, 4.3 and 4.4).

References

Ali-Toudert F (2007) Sustainability and human comfort at urban level: evaluation and design guidelines. In: Portugal SB 2007 conference—sustainable construction, materials and practices: challenge of the industry for the New Millennium

Andreucci MB (2017) Progettare green infrastructure. Tecnologie, valori e strumenti per la resilienza urbana, Wolters Kluwer Italia, Milan

Antonovsky A (1979) Health. Jossey-Bass Publishers, Stress and Coping. San Francisco

Arcidiacono A, Ronchi S (2021) Challenges for contemporary spatial planning in Italy. Towards a new paradigm. In: Arcidiacono A, Ronchi S (eds) Ecosystem services and green infrastructure. Cities and nature. Springer, Cham. https://doi.org/10.1007/978-3-030-54345-7_1

ARUP (2014) City alive. Rethinking green infrastructures, London

ARUP (2016) City alive. Towards a walking world, London

Babic N (2021) Superblocks—the future of walkability in cities?. Academia Letters, Article 747, https://doi.org/10.20935/AL747

Balletto G (2022) Some reflections between city form and mobility. Dilemma between past and present. TeMA J Special issue 1(2022) new Scenario for safe mobility in urban areas, p 7–15. https://doi.org/10.6092/1970-9870/8651

Bartoloni M (2021) Case di comunità, ecco l'identikit: servizi h24 con medici e infermieri. IlSole24Ore, 31 October 2021, retrieved from https://www.ilsole24ore.com/art/case-comunita-ecco-l-identikit-servizi-h24-medici-e-infermieri-AEPGi2r

Barton H, Grant M (2006) A health map for the local human habitat. J Royal Soc Promot Public Health 126(6):252–261. https://doi.org/10.1177/1466424006070466

Beck U (2018) La società globale del rischio, Asterios

Bibri SE, Krogstie J (2020) Smart eco-city strategies and solutions for sustainability: the cases of royal seaport, Stockholm, and Western Harbor, Malmö, Sweden. Urban Science, no. 4, 11. https://doi.org/10.3390/urbansci4010011

Blečić I, Cecchini A, Fancello G, Talu V, Trunfio GA (2015) Camminabilità e capacità urbane: valutazione e supporto alla decisione e alla pianificazione urbanistica, Agenzia delle entrate. 10.14609/Ti_1_15_4i

Borlini B (2010) Il quartiere nella città contemporanea. Come e perché occuparsene. Quaderni di Sociologia, no. 52/2010, OpenEdition J 13–29. https://doi.org/10.4000/qds.717

Brown K, Mijic DA (2019) Integrating green and blue spaces into our cities: making it happen. Briefing paper no. 30, Imperial College London

Browne M, Allen J (2011) Enhancing the sustainability of urban freight transport and logistic. Transport and communication bulletin for Asia and the Pacific, no. 80, retrieved from https://www.unescap.org/sites/default/files/bulletin80_Article-1.pdf

Campos Venuti G (1983) Città, metropoli, tecnologie. Franco Angeli, Milan

Campos Venuti G (2004) Verde in città per urbanistica e ambientalismo oggi. Relazione al Convegno 'Una città per il verde', Padova, February 2004. Retrieved from https://www.pausania.it/per-un-ecologia-applicata-nel-verde/

Campos Venuti G (2009) Ambiente e nuova urbanistica a Modena negli anni Sessanta. In: Bulgarelli V, Mazzeri C (eds) La città e l'ambiente, pp 67–74. APM edizioni, Carpi (MO)

Capolongo S, Buffoli M, Brambilla A, Rebecchi A (2020) Strategie urbane di pianificazione e progettazione in salute, per migliorare la qualità e l'attrattività dei luoghi. Techne, no. 19, Firenze University Press, pp 271–279. https://doi.org/10.13128/techne-7837

Capuano A (2020) STREETSCAPE. Strade vitali, reti della mobilità sostenibile, vie verdi. Quodlibet, Macerata

CEC—Commission of the European Communities (2009) L'adattamento ai cambiamenti climatici: verso un quadro d'azione europeo (COM(2009) 147 def.)

Census (2010) 44° Rapporto annuale sulla situazione sociale del Paese. Il sistema di welfare, pp 257–348. Retrieved from https://www.censis.it/sites/default/files/downloads/welfare.pdf

Cerasoli M (2015) Qualità urbana, mobilità, qualità della vita: una 'grammatica' per il Rinascimento della città. Urbanistica Informazioni, 263:16–19

Chiesura A (2018) Infrastrutture verdi. Qualità dell'ambiente urbano—XIV Rapporto, ISPRA Stato dell'Ambiente, no. 82

Dell'Aira PV (2017) La superficie dello spazio pubblico. Il lavoro 'sul piano' nel progetto dei vuoti urbani. L'ADC, L'architettura delle città. J Sci Soc Ludovico Quaroni, no. 8/2017, Rome

Deloitte (2020) Outlook Salute Italia 2021. Il Sistema Sanitario Italiano tra pubblico e privato: sostenibilità e prospettive, 22 January 2020, Rome. Retrieved from https://www.deloitte.com/content/dam/Deloitte/it/Documents/public-sector/Deloitte%20%Outlook%20Salute%20Italia%202021_Presentazione%20risultati% 20ricerca.pdf

Donati A (2017) Un passo dopo l'altro: nascita e crescita delle aree pedonali in Italia. 'Mobilità pedonale in città'—XIII RAU (2017). ISPRA Stato dell'Ambiente, no. 75

Dorato E (2020) Preventive urbanism. The role of health in designing active cities, Quodlibet Studio, Macerata

Dovey K, Pafka E (2020) What is walkability? The Urban DMA. Urban Studies 57(1):93–108. https://doi.org/10.1177/0042098018819727

EC—European Commission: Infrastrutture verdi (2013) Rafforzare il capitale naturale in Europa (SWD(2013) 155 final)

EEA—European Environmental Agency (2021) Urban sustainability in Europe—opportunities for challenging times, Briefing no. 3/2021, June 12. Retrieved from https://www.eea.europa.eu/publications/urban-sustainability-in-europe/urban-sustainability-in-europe

Finka M, Ondrejička V, Jamečný L (2016) Urban safety as spatial quality in smart cities. Smart city 360° first EAI international summit, smart city 360°, Bratislava, Slovakia and Toronto, Canada, 13–16 October 2015, Springer Professional

Fior M, Galuzzi P, Vitillo P (2022a) New Milan metro-line M4. From infrastructural project to design scenario enabling urban resilience. Trans Res Procedia 60:306–313. https://doi.org/10.1016/j.trpro.2021.12.040

Fior M, Galuzzi P, Vitillo P (2022b) Well-being, greenery, and active mobility. Urban design proposal for a network of proximity hubs along the new metro line in Milan. TeMA Special Issue (1) new scenarios for safe mobility in urban areas, pp 17–30. https://doi.org/10.6092/1970-9870/8650

Gasparrini C (2015) Nella città, sulla città—in the city, on the cities, List, Trento

Gasparrini C (2017) Una buona urbanistica per convivere con i rischi. Urbanistica, no. 159/2017, pp 4–9, INU Edizioni, Rome

Gasparrini C, Terracciano A (2017) Dross city. Metabolismo urbano e progetto di riciclo dei drosscape, List, Trento

Geddes da Filicaia M (2020) La sanità ai tempi del coronavirus. Il Pensiero Scientifico Editore, Rome

Gehl Architects (2020) Public space and public life during Covid-19. Retrieved from https://covid19.gehlpeople.com/lockdown

Guzzabocca C, Legoratti S (2021) Una nuova prossimità. La riappropriazione della strada ai tempi della pandemia. Master's Thesis in Progettazione dell'Architettura, Politecnico di Milano, A Y 2020–2021, Supervisor Prof. M Buffoli, Co-Supervisor Prof. M Fior, Milan.

ILO—International Labour Office, UNICEF—United Nations Children's Fund (2021) Child labour: global estimates 2020, trends and the road forward, ILO and UNICEF, New York

Indini P (2015) La strada come progetto di spazio pubblico. Strumenti convenzionali e non convenzionali per il progetto dello spazio pubblico a partire dalle differenze degli individui. Ph.D.'s Thesis in Architettura e Pianificazione XXVII Ciclo, A.Y. 2014/2015 at Università degli Studi di Sassari, retrieved from Street as project of public space—La strada come progetto di spazio pubblico by Paola Indini

ISCA—International Sport and Culture Association (2015) The economic cost of physical inactivity in Europe. Centre Econom Bus Res

ISTAT—Istituto Nazionale di Statistica (2019) Conoscere il mondo della disabilità: persone, relazioni e istituzioni. Retrieved from https://www.istat.it/it/archivio/236301

John H, Marrs C, Neubert M (ed) (2019) Manuale sulle Infrastrutture Verdi—Basi teoretiche e concettuali, termini e definizioni, estratto in italiano. Interreg Central Europe MaGICLandscapes Project, Dresda. Retrieved from https://www.interreg-central.eu/Content.Node/MaGICLandsca pes-Manuale-sulle-Infrastrutture-Verdi.pdf

Kohl HW 3rd, Craig CL, Lambert EV, Inoue S, Alkandari JR, Leetongin G, Kahlmeier S (2012) The pandemic of physical inactivity: global action for public health. In: Lancet physical activity series working group 380(9838):294–305. https://doi.org/10.1016/S0140-6736(12)60898-8

Kurcharski A (2017) Il futuro della medicina. In: Al-Khalili J (ed) Il futuro che verrà. Quello che gli scienziati possono prevedere, pp 55–65. Bollati Boringhieri, Turin

Legambiente (2020) Ecosistema urbano. Rapporto sulle performance ambientali delle città 2020, In Laurenti M and Bono L (eds), Legambiente

Legambiente (2021) Ecosistema urbano. Rapporto sulle performance ambientali delle città 2021, In Laurenti M and Trentin M (eds), Legambiente

Lynch K (1960) The image of the city. It tran 2006, L'immagine della città, Marsilio, Venice

Manigrasso M (2019) La città adattiva. Il grado zero dell'urban design, Quodlibet, Macerata

Manzini E (2021) Abitare la prossimità: Idee per la città dei 15 minuti, Egea

Marchigiani E, Basso S, Di Biagi P (2017) Esperienze urbane. EUT Edizioni Università di Trieste, Trieste, Spazi pubblici e città contemporanea

Merlo G (2015) L'attrazione speciale. Maggioli, Santarcangelo di Romagna

Miano P (2020) HEALTHSCAPE. Nodi di salubrità, attrattori urbani, architetture per la cura. Quodlibet, Macerta

Morello E (2021) Città visibili: lo spazio urbano come luogo di cura allargato. Fondazione Giangiacomo Feltrinelli. Retrieved from https://fondazionefeltrinelli.it/citta-visibili-lo-spazio-urbano-come-luogo-di-cura-allargato/

Mosca EI, Capolongo S (2020) A universal design based framework to assess usability and inclusion of buildings. In: Gervasi O et al (eds) Computational science and its applications—ICCSA 2020. Lecture Notes in Computer Science, vol 12253, pp 316–331. Springer, Cham. https://doi.org/10.1007/978-3-030-58814-4_22

Nomisma (2021) PNRR missione salute. Verso una sanità territoriale: un piano di investimenti in strutture. Report July 2021. Retrieved from https://www.rekeep.com/media/studi-dossier/nuova-sanita-di-prossimita

Oliva F ()1993 Urbanistica ed ecologia. In: Campos Venuti G, Oliva F (eds) Cinquant'anni di urbanistica in Italia. 1942–1992, pp 201–219, Laterza, Bari-Rome

Panzini F (1993) Per i piaceri del popolo. L'evoluzione del giardino pubblico in Europa dalle origini al XX secolo, Zanichelli, Bologna

Quammen D (2012) Spillover. Animal infections and the next human pandemic, It. tran., 2020, Spillover. L'evoluzione delle pandemie, Adelphi, Milan

Raganà C (2016) Storia della pedonalizzazione: da dove vengono le zone pedonali. Tuttogreen. 7 May 2016. Retrieved from https://www.tuttogreen.it/pedonalizzazione-citta/

Ramboll (2016) Strengthening blue-green infrastructure in our cities. Enhancing blue-green infrastructure and social performance in high density urban environment. Ramboll Foundation

Ravagnan C, Cerasoli M, Amato C (2022) Post-Covid cities and mobility. A proposal for an antifragile strategy in Rome. TeMA J Special issue 1(2022) new Scenario for safe mobility in urban areas, pp 87–100. https://doi.org/10.6092/1970-9870/8652

Ricci L (2021) Città contemporanea e nuovo welfare. Una rete di reti per rigenerare la città esistente. In: Poli I.: Città esistente e rigenerazione urbana. Per una integrazione tra Urbs e Civitas, pp 11–27. Aracne

Ristvej J, Lacinák M, Ondrejka R (2020) On smart city and safe city concepts. Mobile Netw Appl 25:836–845. https://doi.org/10.1007/s11036-020-01524-4

Ronchi S, Arcidiacono A (2021) Lessons from Italian experiences: bottlenecks, new challenges and opportunities. In: Arcidiacono A, Ronchi S (eds) Ecosystem services and green infrastructure. Cities and Nature. Springer, Cham. https://doi.org/10.1007/978-3-030-54345-7_17

Rossi I (2022) Paris 2024, plan of accessibility of roads and public spaces. Urbanistica, no. 164/2022, INU Edizioni, Rome

Sallis J F, Cerin E, Conway T L, Adams M A, Frank L D, Pratt M, Salvo D, Schipperijn J, Smith G, Cain K L, Davey R, Kerr J, Lai P C, Mitàs J, Reis R, Sarmineto O L, Schofield G, Troelsen J, Van Dyck D, De Bourdeaudhuij I, Owen N (2016) Physical activity in relation to urban environments in 14 cities worldwide: A cross-sectional study. May 28(387), pp. 2207–2217. Lancet https://doi.org/10.1016/S0140-6736(15)01284-2.

Secchi B (1986) Progetto di suolo/Projects for the ground. Casabella, 520–521:19–23

Secchi B (2011) La nuova questione urbana: ambiente, mobilità e disuguaglianze sociali. CRIOS 1(2011):89–99

Secchi B (1989) Lo spessore della strada. Casabella, no. 553–554, January, p 38

Secchi B (2006) Progetto di suolo 2. Spazi pubblici contemporanei. In: Aymonino A, Mosco VP (eds) Spazi pubblici contemporanei. Architettura a volume zero, Skira, pp 287–291

Secchi B (2013) La città dei ricchi e la città dei poveri, Laterza, Rome-Bari

Sharifi A (2019) Resilient urban forms: a review of literature on streets and street networks. Build Environ 147:171–187. https://doi.org/10.1016/j.buildenv.2018.09.040

Spirito P (2013) La nuova stagione delle pedonalizzazioni. Huffington Post, 26 October 2013. Retrieved from https://www.huffingtonpost.it/pietro-spirito/la-nuova-stagione-delle-ped onalizzazioni_b_3815731.html

Stefanelli T, Di Bartolo C, Galli G, Pastori E, Quak H (2020) Policy note. Smart choices for cities. Making urban freight logistics more sustainable, Civitas Wiki Consortium

Tamminga K, Cortesão J, Bakx M (2020) Convivial greenstreets: a concept for climate-responsive urban design. Sustainability 12:3790. https://doi.org/10.3390/su12093790

Toppetti F, Ferretti LV (2020) La cura della città. Politiche e progetti, Quodlibet, Macerata

UN-Habitat—United Nation Habitat (2020) UN-Habitat guidance on Covid-19 and public space

UN-Habitat—United Nation Habitat (2021) Cities and pandemics: towards a more just, green and healthy future. United Nations Human Settlements Programme, Revised edition

Urbanistica n. 164/2022, 'Città Accessibili', INU Edizioni, Rome (2022)

Vich G, Marquet O, Miralles-Guasc C (2019) Green streetscape and walking: exploring active mobility patterns in dense and compact cities. J Transp Health 12:50–59. https://doi.org/10.1016/j.jth.2018.11.003

WBG—World Bank Group (2020) Poverty and shared prosperity 2020, international bank for reconstruction and development. The World Bank, Washington DC

WEF—World Economic Forum (2020) The future of the last-mile ecosystem. Transition roadmaps for public- and private-sector players, world economic forum, Cologny-Geneva

WHO—World Health Organization (2010) Global recommendations on physical activity for health. WHO Press, Geneva

WHO—World Health Organization (2018) Global action plan on physical activity 2018–2030. More active people for a healthier world. WHO Press, Geneva

WHO—World Health Organization (2021) Green and blue spaces and mental health. New evidence and perspectives for action. WHO Press, Geneva

WHO—World Health Organization (2009) Global health risk. Mortality and burden of disease attributable to selected major risks. WHO Press, Geneva

Wilson EO (1984) Biophilia. Harvard University Press, Cambridge

Zheng C, Huang WY, Sheridan S, Sit CH, Chen XK, Wong SH (2020) Covid-19 pandemic brings a sedentary lifestyle in young adults: a cross-sectional and longitudinal study. Int J Environ Res Public Health 17(17):6035. https://doi.org/10.3390/ijerph17176035

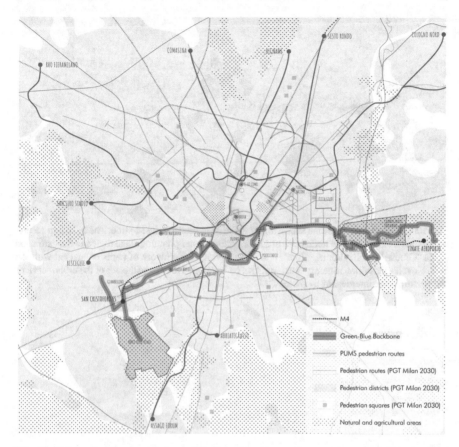

Fig. 4.1 Schematic diagram of the pedestrian program for Milan and the potentialities of the Green–Blue Backbone

Fig. 4.2 Analyses of pedestrian accessibility to the 21 stations of the M4 line. The 5, 10 and 15 min isochrones from the station node were used during the master planning stage to define the extension of the 'Green–Blue Backbone effect', that is, to identify the network of paths connected with the main bicycle-pedestrian itinerary (east–west) to be developed to expand the use of the urban services and performance intercepted by the new M4

Fig. 4.3 Diagrams of
pedestrian accessibility to
the stations of the M4 line.
Defining the mobility
interventions required to
increase pedestrian access to
the subway stations, the
connections between diverse
parts of the city
(centre/peripheries) and
between the
attractors/generators of users
made it possible to promote
variable speed itineraries
(rapid by subway, slow by
bicycle)

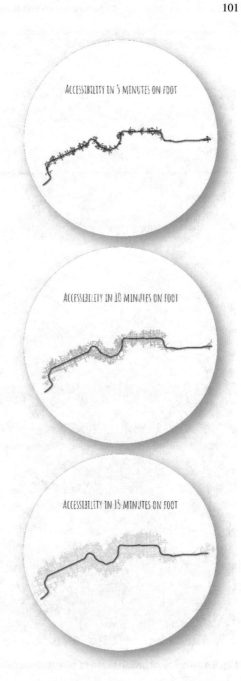

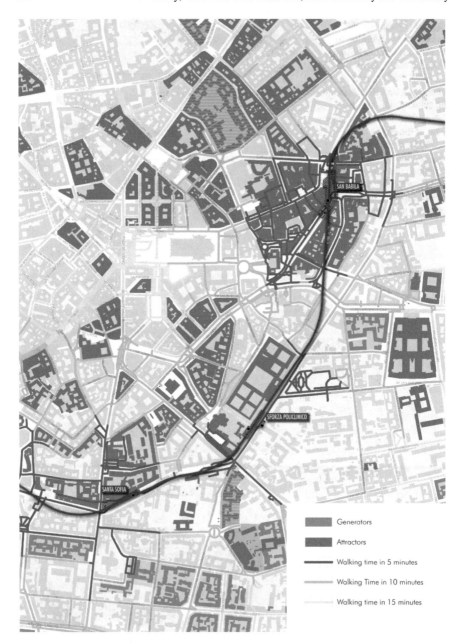

Fig. 4.4 Analytical investigation of pedestrian accessibility to the Sforza-Policlinico station

Chapter 5
Urbanism in Action

Abstract The proposal of urbanism in action accompanies the virtuous local processes that enhance the programmes, projects, and actions to maximize the overall and capillary regeneration processes of the contemporary city. This starts with an awareness of the importance of a multiscalar approach, the constitutive dimension of urban phenomena. A dimension specified by the Neighbourhood Master Plans attentive to the definition of economic and social proximity networks. It is a flexible project, open to different forms and uses, aiming to improve the quality and characteristics of public and collective space. A project that is substantiated by the definition of recommendations and suggestions that will guide its subsequent implementation phases. The case study of the M4 masterplan is a true contextualisation project for the new Milan metro line, characterising it as a blue and green linear park linking two extensive environmental systems on a metropolitan scale.

Urbanism is an action, a practice. It cannot be otherwise; it is knowledge; it is accumulated know-how that, inside the university, fosters an action research directed toward public engagement (*Terza Missione*) and thus attention and awareness of society, communities and territories. How are we to stimulate, favour and accompany projects that promote proximity in local environments and communities? Setting out from this premise and this question, the chapter is organised in four paragraphs. The first proposes a working method that studies multi-scale dimensions and design, from metropolitan projects to urban design to architecture, to create a proximity city that is active, pleasant, civil, rich with differences, open and inclusive. The second part suggests the construction of a neighbourhood Master Plan founded on the reconfiguration of networks and economies of proximity, in which the management over time of shared choices for urban regeneration plays a fundamental role in achieving this aim. The third part listens to and learns from communities and sheds light on the contamination among the technical know-how of experts and local knowledge for improving projects and co-design practice. It is an understanding oriented toward action and enhancing the capacitation of territories, in which processes of active civic participation are both valuable and preferable. The fourth and final part, "The M4 Master Plan, an Enabling Platform for Urban Regeneration", sheds light on ways of working at different scales (from the city to the square) through coherence of

© The Author(s), under exclusive license to Springer Nature Switzerland AG 2022   103
M. Fior et al., *(Re)Discovering Proximity*, PoliMI SpringerBriefs,
https://doi.org/10.1007/978-3-031-08958-9_5

essential operations that help build local knowledge and sharing among urban actors and design issues.

5.1 Working with Multi-Scale Dimensions and Projects

5.1.1 The Contemporary City, a Stratified Palimpsest

Today, contemporary urban design finds itself dealing with the proper form of the city and territories. They differ radically even from those of a recent past: a multi-dimensional and fragmented reality, not only in its physical shape, which modern design was often incapable of dealing with.

This combination of variable geometries, ranging between the spatial, the economic and the social, depends on context and assumes a composite statute at the crossroads between a strategic, spatial and institutional dimension, linked to the design of the most important interventions for planning the city and resolving its most complex problems (Clementi 2004). It is only to facilitate comprehension that we continue to refer to this urban form as the city. This form can no longer be related to an omni-comprehensive observation that expresses a strong and mono-dimensional rationality (Pasqui 2017). Instead, it is a complex system to be governed by a project for frameworks, strategies and tactics (Ippolito 2012; Galuzzi et al. 2019) that hold together material and immaterial actions, exploration, recycling, bricolage (Zucchi 2018), the temporary use of space, open toward interaction with the metabolism of the city and its flows. Thus it is not the shape that defines the contents of design. Still, its relations define the framework that rapidly changes characters, roles, actors and statutes at the urban, metropolitan and regional scale (Amin and Thrift 2016; Balducci et al. 2017). Only within the nature of these relations is it possible to recompose coherence patterns among diverse and heterogeneous materials in their role, form and dimensions.

The question of scale crosses urban studies, from representation to design, describing diverse ways of reading and interpreting the territory, distinguishing among practices and techniques referred to different disciplinary statues, moving beyond the *cliché* that a difference in scale recognise architecture from urbanism. The large scale is not the exclusive domain of planning and programming, quantitative knowledge and a normative design approach. The large scale is not far from or antithetical to the space of perception, the *genius loci* and urban morphology. Undoubtedly, they represent the privileged field of urban design and architecture. The opposition between planning and architecture appears anachronistic. It is centred on the difference between subjects and procedures, resources and times, but above all on effectiveness, in a context in which socio-economic and urban transformation looks to an urban plan to provide a manifest level of operativity for managing changes taking place. The city and the territory are stratifications of levels of inter-pretation at diverse scales, palimpsests, and over-writings; in other words, the results of processes (Corboz 1985). Scale and measure are criteria for interpreting, representing, discretising and recomposing elements and parts in hierarchies or relationships. They are helpful when investigating the physical and social territory, to delineate criticalities and potentialities. Imagining the territory as a multi- and trans-scalar assemblage of heterogeneous and complex materials and phenomena is a means for intersecting the physical, material and visible components of space with the immaterial dimension of those who are the protagonists of space. Moreover, it permits an understanding of space as the result of a mediation between multiple instances— economic, social, and cultural—that, precisely in the name of this complexity, cannot be treated in a segmented form, by separate rationales, by a sector-specific approach (Russo 2011). A multi-scalar condition makes it possible to identify and recognise the foundational materials of the morphologies of the contemporary territory. The multi-scalar approach helps classify territorial forms of mutation caused by rapid, unstable, unpredictable and multidimensional transformation processes. These processes require the growing adaptability of contemporary design. It is possible to consider that scale and measure are indispensable devices for understanding the territory, for correlating the particular with the generic, the detail with the whole. Above all, they establish the importance of the relational aspects of urban and territorial systems as keys to reading their identity, nature, organisation, and regulating principles: crucial factors for identifying their form and structure.

5.1.2 From the Metaphor of the Telescope to the Metaphor of the Mosaic

The 'metaphor of the telescope' (Secchi 2000) was applied to interpret and design cities and territories for many years. It tends to return a linear sequence that ranges from the detailed to the generic. This approach loses specificities, differences and multidimensional relations because it flattens the territory in topographic and two-dimensional representations (Secchi 1989). It moves knowledge away from sensitive aspects, made of practices and flows, and away from the urban populations that belong to and inhabit the diverse layers of the contemporary city.

The 'metaphor of the mosaic' better interprets the contemporary city. It describes and interlocks different dimensions in the multi-scalar and inter-sectorial relations that characterise the structure of the territory, conventionally treated according to sector-specific logics and rationales (Russo 2011), referring to materials distinct by dimension, quality, and grain. At each scale, it is possible to comprehend one of the possible forms of the aggregation and functioning of diverse materials, while the relationship among various scales reveals their specificities (Gregotti 1985). Similarly, actions of soil protection, the mitigation/adaptation of natural and anthropic risks, the conservation of natural resources (water, soil, energy) and the enhancement of historical heritage risk further fragmenting ecosystems if they are not included in a unitary and comprehensive vision. While in urban design a multi-scale condition permits us to look at physical space as the interaction between place and society. Together with the landscape, it can be understood as a way of intending urban design (Gasparrini 2012; Zucchi 2020). The multi-scale condition is the methodological reference for transforming the territory and its systems, its structural components, eco-landscape networks, and identity of place.

A multi-scale condition represents the character and the constituent dimension of urban phenomena. It permits the decodification and representation of the new demands of space. It allows us to confront the problematic node of the forms, uses and powers that represent the specific connotation of the contemporary city. It can be articulated in different dimensions: cognitive and design-based, structural and strategic and political-institutional.

A multi-scale condition delineates a cognitive and simultaneously design-based approach that assumes particular relevance when treating the question of contemporary urban design. It also defines an open and plural field of research for constructing images and visions for the future firmly rooted in the sum of values of territories and specific conditions of a place. The multi-scale condition is a lens for reading urban phenomena and a tool for constructing forecasts. For this reason, the interpretation of a multi-scale condition becomes an inevitable reference for rethinking contemporary urban design (Russo 2011). It is also essential to understand the city, capable of using robust frameworks of knowledge, which cross diverse scales, to capture the measure, dimension, form, and nature of the phenomena mutating the contemporary territory.

5.1.3 A Structural and Strategic Dimension

Scale and dimension define the nature of urban phenomena as well as potential strategies of intervention. They require an integration of multidimensional and trans-scalar strategies, for the long-term, together with coherent, detailed choices (Steinitz 2012). Strategies and actions translate into urban figures and the forms of a structural matrix. The term multi-scale refers to a territorialised method fostering interaction between various measures/actions and diverse dimensions, from society to diverse lifestyles, from the configuration of space to the structure of the landscape. There is a multi-scalar dimension behind programming and the structural tools of spatial planning in current regional legislative frameworks (Belli and Mesolella 2008). In particular, a multi-scale condition is a distinctive character of the ecological-environmental and landscape dimension. This category of interpretation helps treat the complexity of urban phenomena that show an increasing intersection with ecological and environmental topics and with the palimpsests of the landscape, intended as that system of values fundamental to the contemporary design of the territory.

The complex and multisystemic network of Green and Blue Infrastructures may represent the new framework of the contemporary city (Arcidiacono et al. 2018; Gasparrini and Terracciano 2017). It can enliven the built environment and regenerate the consolidated city, focusing on improved performance standards (environmental, infrastructural, settlement). Additionally, the ecological-environmental and landscape framework can intercept the fragmented archipelago of green spaces to reconstruct an ecological continuity with interventions designed to reconnect and reconfigure the environment and redefine a new relationship of coexistence among water bodies and communities, and cities. Green and Blue Infrastructures return to the cross-scale approach of the European project (Estreguil et al. 2019) aimed at constructing robust territorial frames (ecological-environmental and landscape). Similarly, the design of mobility infrastructures implies a twofold operative dimension: on the one hand, a system that identifies selective and broad-scale relations with territorial effects concerning accessibility; on the other hand, that of place, the space of proximity and contact with the local context, from which relations with urban settlements and uses branch out.

In a project for the contemporary city, green space is more than a number, surface or function. It is a complex device: ecological, landscape-environmental and social and health and safety-related: Green and Blue Infrastructures can represent the multi-scale framework of the contemporary city and its resilient metamorphosis.

The time has come to prefigure a comprehensive and integrated project for territories, reimagining contemporary cities and environments together, triggering a green shift toward models of development without growth and centred on the quality of dwelling.

This is possible, in particular, through balanced processes of urban transition, beginning with the role of green space and nature in the city, structured through the definition of multi-scale frameworks: green networks and nodes, at the scale of the territory, at the scale of the city, at the scale of the quarter, at the scale/dimension of proximity; processes that must inevitably be multi-level, multi-actor and multi-sector.

Environmental design, the design of open spaces, of ecological networks, thus represents the primary key to reading current urban and spatial planning policies. Open spaces must pass from being 'design material' to a 'structuring element' of metropolitan landscapes. This can be achieved by leveraging a renewed multifunctionality of services and their rooting in the territory, targeting the welfare and well-being of its inhabitants. Moreover, constructing solid networks of social and collective infrastructures where the system of activities and businesses function as social actors, overcoming the mere logic of economics and profit.

A project for open spaces is naturally trans-scalar, intersecting multiple scales that any good designer must be able to manage. In particular, the system of green spaces that, as we have learned from ecological-environmental disciplines, must be designed as a linked network of points, lines and surfaces to be connected and interconnected at diverse scales: from the extended networks of large green systems (peri-urban and urban); the intermediate networks of neighbourhood parks; the short networks of proximity green spaces (at the scale of the city block/quarter). However, all of this must be done with great humility, abandoning the manias and presumptions of control, often pursued through a detailed and hyper-deterministic project that characterised architectural culture and the modernist approach to green space. Beginning with the configuration of temporary and incremental use, easily reversible and implemented over time based on effective use by different urban populations.

5.1.4 An Institutional Dimension

A multi-scale approach surpasses administrative boundaries, including a plurality of government levels, institutional parties, and stakeholders with different interests. This administrative dimension requires a new dialogue among those public bodies with diverse responsibilities and rationales. Numerous scales and dimensions need other spaces of study and design, involving multiple stakeholders, together with complex institutional interdependencies. Public bodies and institutions complete the social construction of political and administrative choices; using bottom-up processes to pursue issues, programmes and projects, seeking consensus and shared acceptance. The imposition of top-down restrictions on planning is less effective. A shared approach involving stakeholders and institutions helps identify non-negotiable spatial features, which guide the behaviour of a plurality of parties with the constituent characters of the territory. Landscapes, historical fabrics, heritage, environments, biodiversity and infrastructural networks are analysed and interpreted in overall visions and specific projects in a process that crosses multiple scales.

A multi-scale condition is not linked exclusively to physical dimensions and the measure of the territory. Still, it affirms the need to provide a space for decision-making at different levels (governance). Choices can be more durable, permanent, and structural; in other words, they can acquire a more definite, punctual, circumscribed, and temporary character, linked to local and contingent programmes. In this sense, 'scalar relations' concern spatial management devices, urban planning forms and processes and regulating choices. Planning choices have a performance-based and non-regulative character at the large scale, to guide development in proximity spaces between communities and the territory. Vice versa, the criticalities and problems faced by local communities require specific transformations that involve a much smaller scale and interest in managing localised, discrete and molecular actions.

5.2 Neighbourhood Master Plans

5.2.1 Constructing Contemporary Urban Master Plans

Today, urban design must deal with the proper form of the contemporary city. The city is a multidimensional and fragmented reality, not only in its physical configuration. It is an inextricable node of mobility, flows and day-to-day practices. We deal with urban landscapes legible in the 'metaphor of the mosaic': activities and environments in continuous mutation recomposed like mosaic tiles. A sequence of hybrid systems that fuse nature with culture, the rural with the urban, agriculture with industry, and whose reciprocal influence can be recognised solely in leaps in scale (down-scaling/up-scaling). Therefore, contemporary urban design is forced to face issues, problems, scales and dimensions different from those of the past, damningly complicated and impossible to resolve using the tools in the consoling toolbox of modernity (e.g., land use regulation). Hence, we are forced to guarantee the maximum flexibility of interventions over time, convincing processes, moving in diverse directions, from integrating possible uses to evaluating the feasibility of transformations. As a platform of development and inclusion, the city offers the backdrop against which we can work with urban devices, actions, and policies centred on a multi-scale condition. The Master Plan, configured as the tool for exploring the forecasts of a land use plan at the scale of the neighbourhood, appears more suitable, more flexible and adaptable to the themes to be confronted at different scales (Ardielli 2013). It is an urban planning tool, with a strong direction by the public sector, capable of guiding successive programmes and projects, both public and private, in the city. The Master Plan contains a project for the town to be discussed and agreed upon among public authorities and stakeholders. A flexible project, open to diverse forms and uses, and not a final design; a project to improve the quality and characteristics of public space, defining recommendations and suggestions that will guide the actions of interventions that belong to the following detailed projects. As a design tool, it expands the field of interventions to relational aspects, for example, looking not exclusively at housing but also the quality of dwelling in its entirety. This vast theme considers the relationship between built and open spaces, how buildings touch the ground and the hinge function between private and public space.

The above conditions represent the starting point. The Master Plan is a planning device that identifies and searches above all for relations and performance characteristics of the contemporary city. It looks at both process and output, beginning with thematizations of public space. This latter operation holds together different topics (landscape, environmental, infrastructural, and settlement-related) through a project for implementing urban regeneration policies at different levels (from the urban vacuum to the urban block to the building). Urban regeneration must be accompanied by a straightforward public project (on the performance of space) and a transparent approach to managing public–private partnerships. Furthermore, urban regeneration must be confronted through a renewed approach to regulation. The public sector plays a central role, and which rethinks the urban plan as a tool for enabling any perspective intent on effectively and soundly improving the quality of dwelling. For this reason, urban regeneration strategies and actions require a design framework that regulates what is necessary and explains desired performance levels, simultaneously incorporating the dimensions of time and economic feasibility (Lavorato et al. 2020). The awareness of the progressive distance between urbanism and the plural forms of peoples' lives has generated a loss in ties with the concrete experience of dwelling. However, society and its economic forces can be guided and directed during the transformation of the territory, establishing a playing field (its 'lines of force,' the rules of the game) rather than unpredictable outcomes. The objective is to facilitate social aggregations, which do not presuppose sharing horizons of meaning and belonging to a community. In other words, the definition of a reference structure for spatial transformations can support the coexistence of a plural society in the absence of a common background, which remains reversible. It is a question of *aggregazioni compossibili* ('co-possible aggregations,' situations possible simultaneously) open to variety, differences, eventuality of the possible, uncertainty, unpredictability, and risks (Pasqui and Sini 2020). This approach thinks of space as the effect produced by the operations that orient it, circumstantiate it, time it and make it function. Space is thought of as a polyvalent unit of conflictual programmes or 'contractual proximities' (de Certeau 1980).

The contemporary urban Master Plan is the search for a 'figure' that gives meaning, measure and order to architecture (Choay 2003; Mei 2015): observing and distinguishing objects means organising a figure in the foreground and shifting the rest into the background. We must not design a space to define a figure; we design it because different figures correspond to different uses, practices, and relations. It consents local systems to adapt to changes, creating environments in which forms of governance, typologies of services, social mix, economic devices and mechanisms can cooperate so that communities can adapt, also in relation to the physical aspects of dwelling. By radically focusing on the needs and expectations of citizens (Consonni 2019), who, with their day-to-day lives, occasional and unpredictable, interact to define space and give form to the city.

5.2.2 Constructing Networks of Proximity

Urban figures represent manifest intentions, sequences of spaces, concatenations of places that return to the concept of the structure and the frame: rooting design to space and through political strategies and local actions. Achieving this requires working by layers/systems through a process of selection and hierarchisation aimed at constructing 'networks of proximity.' The new design perspective involves defining a system comprised of places, routes and pedestrian and cycle paths, green areas and accessible and inhabitable urban spaces. The interest of urban studies must be shifted toward observing the diversity of practices and uses in space and a qualitative and quantitative relationship with the different scales of the contemporary city (territorial, municipal, local).

Therefore, form cannot be a starting point for design, nor can it be the output of a norm. It must be the result of a process to be managed with the necessary flexibility; no project can be abstracted from its real and specific conditions but must propose operative methods. Additionally, every project raises the question of identity: building identity is both a process and an opportunity. We can assume urban design as:

- searching for the most virtuous correspondence between the characters of places and functional vocations,
- guiding and orienting transformations through the definition and emphasis on existing 'lines of force,' with a strategy for adapting to context often ignored by modernist design;
- working patiently with the context features to construct solutions based on and motivated by its understanding and interpretation.

Understood in this manner, public space becomes a 'generator,' the binder in a sequence of different environments, articulated in a close relationship with the fabric that delimits it, identifying a pattern capable of giving meaning to open spaces.

In the end, we can consider 'proximity' as an integrated system of neighbourhood (public) spaces made of green and mineral open spaces. With the construction of proximity, the common ground, innervated and revitalised by the network system, is historically the ecosystem of urban socio-economic development. The common ground of the city is the space where different activities and uses co-penetrate one another, and the fundamental prerequisite for guaranteeing an urban mix. This space is home to developments in which the bases of buildings are occupied by 'economies of proximity' (small shops, artisanal services, public establishments), whose importance has been rediscovered: the baker, the fruit seller, the grocer, the newsstand, that once animated neighbourhood life and seemed like useless activities in a contemporary era of consumerism and metropolitan lifestyles. These activities foster exchanges with the city, animate the common ground of the neighbourhood and become spaces of urban encounters, exchanges, interactions and permeability. Each place welcomes different practices of meeting and pausing, and attracts flows of people, oriented by the need to carry out necessary or voluntary activities (Gehl 1971). This construction of proximity will permit a renewed continuity and new typologies of spaces and uses: porous lots, thanks also the presence of vegetation inside courts and courtyards, make it possible to define a new system of urban relations between private and public spaces and to favour actions of crossing. By learning from the historical city, this contemporary urbanity mixes different though strongly integrated activities and functions that dialogue with the system of open spaces. It defines a constellation of uses and activities to be promoted, as per Milan's first PGT (Arcidiacono et al. 2013), recognising their social and collective value, the role of vitality, and at the same time, their role as an urban outpost. New accessibility is made of juxtapositions and overlaps of diverse networks, needs, and practices. Within this framework, the space between objects demonstrates a capacity to assume the same importance as the objects themselves. The new design perspective extends public space creating a new network of shared and accessible areas, improving the quality of dwelling, and developing a sense of belonging to the neighbourhood. The re-configuration of open public and flexible private spaces, available for different uses, also offers new everyday activities and practices.

Proximity is reimagined as richness with a social and economic value. The Covid-19 pandemic has generated a great opportunity: to radically reform our homes (our shelter, our intimate and domestic space) squashed by real estate standards indifferent to living conditions and lifestyles. Moreover, the pandemic reforms our cities (where we dwell) and our existence in the world. The objective is to construct a system of settlement articulated to build an organic and integrated archipelago of strongly related neighbourhoods that privileges public mobility and pedestrian and bicycle connections, comprised of different neighbourhoods, each with their own identity within a simple urban composition. The aim is to create articulated centralities, true and proper cores for neighbourhood life that foster the networking of urban spaces and the improvement of diffuse polarities. This means avoiding the ordinary periphery, preferring true and proper pieces of the city, tiles in a mosaic with a specific identity, and promoting services and local activities rediscovered during the pandemic.

5.2.3   Placing Slow Mobility at the Centre of Urban Life

In the early 1970s, another global crisis, in this case energy-based, triggered a revolution in the system of transportation, above all in Northern Europe. In particular, the Dutch Government initiated a robust structural and operative programme to construct bicycle paths. Today, some 30% of all travel across the country takes place by bicycle.

Similarly, during their current global crisis, this time caused by a pandemic, there is no shortage of international examples to draw inspiration. Oakland (California), with its 'Slow Streets' initiative, has converted approximately 120 km of city streets for use by pedestrians and cyclists, limiting vehicular traffic to local movements. The New Zealand Government has promoted financing for light tactical urban interventions to create bicycle paths and widen existing pedestrian paths, permitting the latter to respect social distancing norms.

Similar interventions were common in Europe, even before the pandemic. Berlin was already working to widen existing bicycle paths or create new ones to balance the reduction in public transport. In Barcelona, the *Superilles* project is multiplying pedestrian routes through a progressive and straightforward hierarchisation of traffic. Paris announced its proposal for a 100% cyclable city by 2024, supporting the idea of a city at the human scale, where all inhabitants will be able to access principal neighbourhood services in less than 15 min (1.5 km).

To diversify the offering of mobility, in the coming years it will be necessary to identify alternative solutions for supporting collective transportation. While the idea of promoting pedestrianisation and cycling in cities was valid in ordinary times, it will become even more critical and strategic during the post-pandemic phase ahead of us. The best means for moving and integrating physical activity in cities is also the simplest: walking and cycling. Walking is an unavoidable necessity and the simplest and most natural form of movement we know, in particular at the human scale of the neighbourhood. The bicycle is perhaps the safest alternative to public transport for moving through the city. It will be essential to favour walking and cycling to lighten the load on local public transport and permit outdoor activities that respect social distancing imposed by the pandemic. It will be necessary to extend the space of sidewalks and create new pedestrian and shared spaces, guaranteeing the safety of those who walk and identifying new ways of imagining public space and socialisation. We must redefine the use of streets and public spaces and develop areas that permit commercial, recreational and cultural uses, and sport. We will have to act quickly to offer an alternative to the automobile in response to the needs of mobility of citizens, guaranteeing safety concerning the measures planned for public transport, offering incentives to active mobility as an alternative and/or integration for movements at the urban territorial scale. The pandemic may represent an opportunity for taking a decisive leap forward toward the more widespread use of all forms of micro-mobility (bicycle, both traditional and assisted, kick/electric scooters). Despite a significant increase in the use of these means in recent years, we are still far from their potential to transform urban mobility.

Some experiments and projects underway in Milan move in this direction, with strategies, actions, instruments, and devices that promote cycling and walking, also as a means of guaranteeing distancing during movements through the city in the wake of the pandemic, but more in general for promoting an integrated system of sustainable mobility (Comune di Milano 2020).

Particular examples include the following action programmes: the *Piano di Governo del Territorio (PGT Milano 2030)*; il *Piano Urbano della Mobilità Sostenibile* (PUMS 2018) and integrated actions to promote emergency cycling itineraries during the pandemic; the *Piazze aperte in ogni quartiere*; the *Strategia di adattamento Milano 2020*.

The *PGT Milano 2030*, part of an articulated system of policies to promote urban regeneration, identifies a network of spaces dedicated to pedestrians. In these areas, interventions are activated to moderate traffic and care for the city, leading to improved quality of life, both environmental and social. The network, conceived as a framework of collective urban life at the centre of neighbourhoods, aims to attract and support the operativity of small enterprises, craft-based and creative activities and the connection of society, culture and community. The principal actions are concentrated in the city's quarters, and in the connection between them, to guarantee access to essential services of proximity, to facilitate the return of commercial activities, also outdoor, and to provide space for movement and cultural and social events, all under the safest conditions.

Several actions have also been implemented to address the pandemic crisis in coherence with the PUMS 2018. To promote effective active mobility, through the realisation of a system of cycling itineraries (a structuring framework), both radial for the connection with the quarters of the city farthest from the centre and with the municipalities of the Metropolitan City; and annular and transversal to favour systematic movements between different urban centralities. Structuring itineraries are integrated with interventions for cyclability and moderated traffic zones (zone 30) for safe and liveable mobility at the level of the quarter. During the pandemic, it became even more important to establish relations between the city's quarters and the Metropolitan City. New emergency cycling itineraries reinforce connections between rapid mass public transport (the metropolitan rail network) and the urban area to offer everyone the possibility to utilise an alternative form of transport to the automobile to reach their workplace. With these aims in mind, the objective was to notably extend and better connect the existing cycling paths by creating approximately 35 km of new routes. Together with this plan, other interventions are already planned to augment the urban cyclability network, destined to further improve the entire network. This strategy accompanies that of the *Città 30* indicated in the PUMS 2018: an increase in 30 km/h zones above all in the quarters of the city outside the bus ring road (*circonvallazione filoviaria*), in part realised with signage and in part with structural elements for moderating speeds, ensuring road safety and widening pedestrian and landscaped spaces. The cycling network connects new and existing zone 30 s, which benefit from cycling itineraries with linear paths created solely using signage and/or simply indicated along 30 km/h streets or streets shared with other vehicles.

Since 2018, the programme *Piazze Aperte in ogni quartiere*, in collaboration with Bloomberg Associates, the National Association of City Transportation Official (NACTO), and the Global Designing Cities Initiatives, has experimented with the method of tactical urbanism to generate new public spaces that substitute redundant streets or intersections, through light, rapid and economic interventions of an experimental nature, prior to being consolidated. Their temporary nature permits rapid action and the testing of reversible solutions, before investing time and resources in the final design, anticipating effects with immediate benefits and supporting the decision-making process that leads toward a permanent solution. These experiences have permitted the development of new skills and the testing of a new kit of tools for intervening in public spaces. 'Light' projects implemented in the city's quarters through various projects of pedestrianisation and the activation of space include the widening of sidewalks and the creation of cycling paths using only signage, protected against car parking by bollards. With the collaboration of associations and citizens, more than 60 proposals were received, currently in the phase of co-design/realisation. One virtuous example is offered by the experiments in Piazza San Luigi, in a quarter in the southern part of the city, beyond the Porta Romana railway backyard. Through a phase of experimentation and co-design, this project worked to requalify the public square in front of the church, eliminating a parking lot and revitalising the system of neighbourhood activities, also within a larger context of reference.

The *Strategia di adattamento Milano 2020* is a document open to contributions from associations and citizens. It proposes rethinking the hours and rhythms of the city to reduce and redistribute the demand for mobility throughout the day, improve and further diversify the offering of mobility, implement and exploit the potential for public transport and mass rapid transport infrastructures such as subways and tram lines. This strategy is paralleled by the promotion of sustainable and active mobility and sharing and redefining the use of streets and public spaces and non-polluting ways of moving across the surface of the city (walking, cycling, and soft mobility). Some 3,000 contributions/proposals have been received for the areas of mobility, the environment, business and urban planning.

5.2.4 Restarting from Nineteenth Century Modernity

The rule used to organise and compose the European city, generated by the Industrial Revolution and nineteenth century modernity, is represented by a system of neighbourhoods whose primary cell is the urban block (Panerai et al. 1981; Sica 1992; Zucconi 2001). In Milan, as in many other European cities (Paris, Berlin, Madrid, Barcelona), a characteristic structure and morphology of settlement produced urban grids with similar and recurring forms, dimensions and granularities. There is a continuous improvement of homogenous 5/6 storey street fronts, lacking any functional specialisations, except for blocks destined for industrial uses. The feet of buildings were 'naturally' home to functions permitting exchanges with the city: shops, artisans' workshops, public and private services. Works of architecture capable of appearing normal, and being politely metabolised within the morphological and functional transformations of contemporary society. This important morphological and material legacy could opportunely represent the physical and material palimpsest atop which to reconstruct a new urbanity framed by interconnected proximity networks.

For nineteenth-century modernity, the daily routine was strictly related to the spaces of the home and workplace, and its rhythms determined by the factory. The successive post-industrial phases, globalisation and its effects pulverised needs, uses and practices, no longer linked to rhythms defined by home-work. Several relevant and pertinent critiques of functionalist theories attempted to overcome some of the rigid and abstract aspects of the Modern Movement. With the purely functionalist conception of the city and the mechanist conception of society, critiques aimed to ensure greater consideration of citizens' needs, the search for a new contextuality for design and a new relationship with history (Jacobs 1961; Lynch 1960; Gehl 1971). These positions already emerged within the Modern Movement itself. It did not negate the dependency between space and society (Welter 2005). They promoted a less rigid, functionalised and codified approach to architectural and urban design to capture the articulation and variety of different social morphologies and activities that characterise the life of the historical city. To comprehend the combination of the diverse daily experiences of urban life, other research introduced the metaphor of 'rhythms' founded on the relationship between the body, urban rhythms, and cities (Bonfiglioli 1995; Mareggi 2011). What places us in relation with our surroundings is, above all, our body (Lefebvre 1976): together with the progressive transformation of natural cycles and the consequent fragmentation of time and space, we have determined individual rhythms linked to day-to-day practices. Therefore, time is configured with respect to a 'polyrhythm,' to the speeds and intensities of routine, work and care (de Certeau 1980; Pasqui 2008): space is defined by practices, and through them, inhabitants confirm it.

The morphology and typology structures of settlement inherited from Nineteenth-Century Modernity are naturally resilient to the ordinary urban metabolism. The legacy of Modernism, its buildings and uses, joined with the sense of naturalness of its common ground, can be interpreted and reused in the redesign of proximity. We can use this to realise a new urbanity tuned to the rhythms of the contemporary city and civic participation processes, welcoming the dimension of the unexpected, differences and variations in time.

5.3   Listening to and Learning from Communities

5.3.1   The Community in the Complexity of Contemporary Society

The past two years of the pandemic, characterised by challenging critical situations for both individuals and society, have led to a rediscovery of the value of solidarity and, in particular, community values. We have learned that the most suitable responses to the different ways in which this emergency was manifest were most effective when they began with the spaces in which we live and initiatives of collective solidarity, more or less organised and structured in specific territories (Censis 2021). Relaunching the values of 'proximity,' intended in its most stringent form as the relationship between communities and inhabitable space, both built and open (Vitillo 2021).

Related concepts and theories are re-proposed, echoing a past that is not so remote for Italy and Europe in terms of time, but very distant in its material conditions. We are talking about the post-war period, particularly the post-war reconstruction phase, whose aspirations and contradictions were condensed and, in some cases, clashed, during the first seven years of the INA-Casa[1] programme (Casciato 2001; Secchi 2001). Above all for Italy, this period coincided with a highly delicate phase for the construction of a culture of urbanism and architecture during the democratic-republican phase, animated by a strong ideal tension in research and debate. Nonetheless, we cannot help but note the weak roots and perhaps consequent approximate operative definition of certain concepts, such as the 'quarter,' the 'neighbourhood unit' or 'community' in the events that followed (Caniglia Rispoli and Signorelli 2001). These cultural and political factors influenced correct critical experimentation, not to mention their more incisive application in constructing answers to the expansion of the modern city.

At a time characterised by solutions to systemic crises and emergencies (health, environmental, climate), there is an increasing need to identify economic and social parameters. The parameters are different from those that have dominated economic and social policies (privileging the market and competitivity), as we rediscover alternative reformist models and ideas (Palermo 2022). Alternatively, we rediscover (positively or negatively) models referable to Adriano Olivetti's way of thinking. Unique and peculiar, it remains an influential reference that we must confront, though avoiding the easy traps of mythography (Olmo 2018). Olivetti's ideas are often evoked for their distinctive and collaborative 'sensitivity' to urban, architectural and territorial dimensions. Notwithstanding this affinity, each rapid and accelerated availability of public EU debt funding calls for a focus on welfare policies that—by providing greater support to schools, infrastructure, research, the environment and digital innovation—neglect territories and the 'communities' for which these policies are intended.

Might it still be fertile to refer to the Olivettian notion of community, seeking the conditions for a more concrete and updated interpretation in our contemporary era, without simplifying the thinking and the path for making it a reality? This would appear to be the direction of the most recent practices in urbanism, working to contrast the fragility of territories and foster urban regeneration. Restitching the social fabric and rebuilding communities are the premises for achieving resilient and innovative solutions for society.

---

[1] The INA-Casa programme was a post-war housing plan promoted by the Italian State to build two million units. It takes its name from the *Istituto Nazionale delle Assicurazioni*, National Insurance Institute, which funded the programme.

The current trend, referable to certain widespread and penetrating positions on 'social design' working with scenarios of urban proximity the, '15-min city' seems to now be convinced that a community cannot be planned as it is a form of society that emerges from a diversity of events. Instead, it seems possible to create a suitable environment within which it can be formed and prosper. The environment most suited to this growth and development appears to be that which offers an 'appropriate proximity' that, more than functional, must be relational, and capable of generating social interaction (Manzini 2021). This model privileges light, open, ephemeral communities intentionally created, by choice, around a project for aggregation that appears to be its trigger. When this project involves some form of interest in and caring for space, we can refer to 'communities of place.' Weaving together proximity, community, and care in a constellation of creative communities, of small groups of active people who bring not only needs but also skills (for making and transforming). These communities necessarily exercise their influence over the restricted environments of contemporary life, such as social streets or social districts. They are similar to highly circumscribed entities that, like enzymes, are considered capable of activating processes at a larger scale.

These manifestations show an awareness that these social forms can be generated spontaneously by the joint actions of different actors (conflictual or collaborative). It follows that they cannot be foreseen, nor can the outcomes be known or predetermined, because they are contributions to a more general generative process.

These considerations are not so distant from the concept of community inherited from Olivetti's ideas. Ideas that include building an open community to be monitored and corrected as it develops, far from the critiques and rigid logic of social engineering. The substantial difference lies in its democratic, social and environmental founding values around which communities are built, and which are used to measure the capacity for resilience and durability of their bonds that, in the most recent hypotheses, appear to be non-essential. While communities appear light, their morphologies and unifying ties appear even more ephemeral. In particular, when they are anchored to limited and highly circumscribed projects or contexts. Perhaps their capacity to affect denser processes and a more profound innovation of society that moves beyond the trends, fashions and flashy contemporary lifestyles are even more shifting. An ability that should emerge as a piece of social innovation, or a concrete trace of community (Bagnasco 1999).

The State and the Market are two modern obsessions placed in crisis by the global world of economic flows, systemic global crises and the health emergency. If the community proposal is topical and configured as a Society that defines itself as an alternative to the State and the Market, perhaps the 'moderate Olivettian heresy' can still give meaning to two accursed words of the twentieth century: community and person. As emphasised by Aldo Bonomi "(…) Utilising the new bolstered concept of community is today for me an attempt to move beyond the founding categories of classical sociology for which the concept of community refers to an organic vision of the social relations present in a given situation, founded on the reciprocal compression of its members and a common belonging, subjectively felt by individuals" (Bonomi 2002).

More specifically, it means creating networks between subjects and territories in globalised post-Fordist economies. The aim is to restore sense and meaning to places, rediscovering a human dimension that is neither too large (of flows) nor too small (of 'resentful and claiming localisms'). Networks must be created where social capital loses competitiveness and becomes a bond, a reproducible relational asset. Networks are created where forms of community, resilience and durability can be attributed openly and reflexively to new intermediate bodies. These intermediate bodies must once again act as a link between society and institutions. They are the places where people live and produce and where decisions are taken, i.e., they are the expression of territorial stakeholders and State government.

Finally, it must be remembered that the community project (even within material, social and economic conditions that are different from those of the past) remains a valid but very delicate and complex weaving between actors, territories and intermediate bodies.

5.3.2 Hybridising Different Know-How

The hybridisation of technical-design know-how (expert knowledge) and local know-how (context-related knowledge) is capable of generating forms of 'usable knowledge' (Lindblom and Cohen 1979), obtained through techniques, methods, and forms of participation with the power to increase the listening capacities of communities. Examples include the activation of local discussions that permit the realisation of 'maps of communities': representations of the territory that offer inhabitants the possibility to describe heritage, landscape, qualities and projects in which they recognise themselves (Clifford et al. 2006). This true and proper cultural project that animates the territory is the result of localised, diffuse and specific knowledge of inhabitants. It sheds light on how the local community sees, perceives, and attributes value to its territory, memories, reality, and what it would like this to be in the future. The construction of 'maps of communities' feeds a collaborative process, aimed at constructing an idea of the future city, useful for coordinating actions through 'scenarios of development,' capable of providing communities a context and a trajectory in which they can recognise themselves. Designing by scenarios represents a valid means of collective interaction (Dupuy 2011; Blečić 2012; Blečić and Cecchini 2016; Wade 2012; Alessandrini 2019); with a fundamental awareness of having to incorporate current global risks within operative dimensions—climate change and natural risks (Galuzzi et al. 2020); triggering forms and methods of local activism. This knowledge assumes all the limits of its performance but orients procedures concerning doing and the possibility for real actions; implementing solid and serviceable strategies focused on generating inhabitable space.

These representations aim at describing the environment in which communities live. From the most plural, composite and articulated point of observation and a dynamic conception of social capital intended therefore as the stable readiness of networks, relations, and knowledge, but also an active and moving context, functional to social interaction (Dupuy 2011; Wade 2012; Alessandrini 2019). This living environment for communities can be put to work and placed in a relationship of tension with the capital present in the territory. The goal is to make place-based projects for local development, for places and people, viewed in the most open, composite and articulated manner. Moreover, the aim is to construct informed projects for the territory, reinvent the sense of one's actions in a non-localist key and oriented toward defining Community-Led Local Development (CLLD) strategies.

Some interesting research has shed light on the importance of techniques and methods of participating in activating and improving the capacitation of the territory, above all when they are capable of boosting the listening capacity of local communities (Bobbio 2004). Processes of bottom-up social innovation and transformation are fundamental resources for managing the complexity of the changes and metamorphoses that accompany the shifts of places. These contexts are witness to a significant densification of criticalities and important opportunities for 'transformative resilience' (Brunetta and Caldarice 2020). Their activation begins with best practices tested in processes for the innovation and capacitation of communities (Comim et al. 2010), capable of 'bouncing forward,' as forecast in the Sustainable Development Goals of the UN's Agenda 2030; but above all with the needs and fragilities expressed by settled communities. The understanding of the day-to-day living conditions of local communities, founded on the real necessities of territories, furthermore represents a robust antidote to fragility (Vitale 2009; Blečić and Cecchini 2016): extracting conditions of daily life from territories, constructing pertinent biographies of settled communities, founded on local necessities, is an element of rehabilitation for these same communities, not an abstract anti-fragility, but built atop the needs of local territories, based on their historical, social and economic specificities.

Processes of active civic participation naturally presuppose an effort to listen and engage in patient dialogue; yet they appear both useful and preferable for at least three reasons:

- they contribute to better contextualising urban regeneration projects based on the knowledge of local desires and projects, rooting them to place, improving and elevating the overall quality of performance;
- as with all methods of empowerment, during the construction of participatory processes people, organisations and communities acquire skills that improve their social and living environment (WHO 2006);
- contrary to common belief, they abbreviate the overall duration of processes transforming the territory, dilating the phase of building awareness and defining projects, reducing not only distances between citizens, communities and institutions, but also bringing a notable increase in the total time required to mature decisions, bolstering legitimacy and efficacy during the phase of implementation. Good governance is probably the best approach to planning uncertainty (Stoker and Chhotray 2009; Allegretti 2020).

5.4   The M4 Master Plan, an Enabling Platform for Urban Regeneration

The second volume of the *Sixth Assessment Report* by the Intergovernmental Panel on Climate Change (IPCC 2022) is the most up-to-date and comprehensive evaluation of the global and local impacts of climate change on ecosystems and biodiversity. Moreover, it reveals the consequences on peoples' wellbeing and the Planet. The report highlights that a 1.5 °C rise in average temperatures would cause an irreversible loss of entire ecosystems and significant food and water shortages, exposing people and the environment to risks they will be unable to adapt to: in synthesis, a scenario of permanent crisis. Road-based transport is one of the principal causes of atmospheric pollution and greenhouse gas emissions: it is the leading source of nitrogen oxide and dioxide emissions and the second source of emissions of particulate and carbon monoxide.

Urban areas, in particular, are responsible for 23% of CO_2 emissions produced by transport; in addition to atmospheric pollution, transport is also a relevant source of costs of urban congestion and acoustic pollution in urban areas.

Given that urban mobility has an impact on both economic growth and the environment, for some years the European Commission has been promoting sustainable urban mobility through strategies capable of favouring a transition toward cleaner and more sustainable transport, for example walking and cycling, the use of public transportation, innovative forms of use and ownership of vehicles (Rupprecht Consult 2019).

Sustainable public transport thus represents a fundamental objective for activating virtuous and sustainable urban environmental policies. Rail-based public transport with its own dedicated space can also be configured as a multidimensional enabling platform for urban regeneration: a system of lines and nodes that forms an interconnected network that significantly amplifies the network effect of rail-based transport, with a plurality of functions in relation to different and practical actions and public policies, in addition to contributing as mentioned to favouring the transition toward more sustainable mobility, under at least three profiles: pluri-dimensional and pluri-functional systems, spaces of everyday life, spaces for innovative uses.

The stations represent above all nodal elements in transportation networks, and their centrality contributes to the proper functioning of urban and territorial systems; the network becomes a source of interconnection and connectivity, with the nodes-station that play a fundamental role in their organisation: switches, exchangers, transformers (Dupuy 1996). There is a pluri-dimensionality to the station system (pick-up station, market and station, work and station), in relation to the functions it hosts and the localisation of opportunities that respond to travellers' needs and those of the quarters the network crosses.

Stations are nodal points within transport systems and spaces of everyday life concerning the demographic, economic-productive and urban-territorial context in which they are located. Yet they exceed the mere functionality of the network to assume a decisive role in urban space. Centred on offer and not on the needs of actors and their articulated and multi-location lifestyles, overcoming a purely utilitarian conception, mobility can also reposition opportunities obtained through increased accessibility (Borlini and Memo 2009). True transit nodes, points of access to the possibilities in a territory, can generate a redistributive effect and mobilise spatial and social capital. The stations can also be configured as urban spaces that promote innovative uses and respond to the diversified needs of increasingly more mobile populations. Those who move and temporarily inhabit urban spaces and contemporary mobility. Stations present the capacity to respond to the demand for participation, interaction and the social inclusion of actors, considered values and not in purely utilitarian terms (Pucci and Vecchio 2019).

The project to contextualise Milan's new M4 metro line has shed light on how it can work at different scales: the city, the square and the system of urban open spaces, pedestrian and bicycle networks. Based on the coherence of essential operations that help build knowledge and sharing among urban actors and themes of design. The state-of-the-art of social capital present in some examples of action research on the issues of Milan's former rail lands and the reuse of large brownfields have revealed the existence of a robust framework of social capital and community practices (Montedoro 2013; Comune di Milano 2017).

The new M4 metro line is configured as a blue and green linear park connecting two sizable environmental systems at the large scale (the Parco Agricolo and the Parco delle Risaie to the south-west, the Grande Forlanini-Lambro-Idroscalo system to east), assuming a trans-scalar metropolitan dimension in the municipalities of the first ring (Segrate, Peschiera Borromeo, Buccinasco, Corsico).

The metro line is accompanied on the surface by the Green and Blue Backbone (the spine of new pedestrian-bicycle paths) that completes the forecasts of the pedestrian itineraries of the PUMS (*Piano Urbano della Mobilità Sostenibile*, Sustainable Urban Mobility Plan). Rather than overlapping the underground route, it links together relevant parts of the city crossing metro stations: environmental (landscaping and water), services and infrastructures, history and memory; that relates/connects them to the stations.

From Linate Airport to the Forlanini train station, via the metro station serving the Forlanini district, from Pratone area (grassland) to Piazza Dateo and Piazza Tricolore, along the axis linking Argonne-Susa-Plebisciti roads; crossing the city centre, from Corso Monforte to the Statale station to the Sforza-Policlinico station; from Corso Italia it reaches the Parco delle Basiliche and from here Vetra station, and on to Corso di Porta Ticinese, to de Amicis to Sant'Ambrogio. Then, it runs along Viale Papiniano to Parco Solari, Via Vespri Siciliani, Via Giambellino, to Via Segneri and San Cristofaro stations; it then crosses the Via Giordani viaduct to reach the Parco delle Risaie and the open spaces of the Parco Sud (Figs. 5.1, 5.2, 5.3, 5.4 and 5.5).

References

Alessandrini G (ed) (2019) Sostenibilità e capability approach. Franco Angeli, Milan
Allegretti G (2020) Ricostruire la partecipazione civica nella nuova normalità. Alcuni indirizzi per una possibile rifondazione. Contesti, Città, territori, progetti. Beyond the pandemic: rethinking cities and territories for a civilisation of care. Special Issue, no. 2/2020, pp178–194. Firenze University Press
Amin A, Thrift T (2016) Seeing like a city. Press Cambridge, Cambridge
Arcidiacono et al (2013) Il Piano Urbanistico di Milano (PGT 2012) | The Milan Town Plan (PGT). Wolters Kluwer Italia, Milan
Arcidiacono A, Ronchi S, Salata S (2018) An ecosystemic approach to green infrastructure de-sign in urban planning. Experiments from Lombardy, Italy. Urbanistica, no. 159, pp 102–114. INU Edizioni, Rome
Ardielli M (2013) Masterplan: né piano né progetto. INU Edizioni, Rome
Bagnasco A (1999) Tracce di comunità. Il Mulino, Bologna
Balducci A, Fedeli V, Curci F (eds) (2017) Oltre la metropoli. L'urbanizzazione regionale in Italia. Guerini e Associati, Milan
Belli A, Mesolella A (eds) (2008) Forme plurime della pianificazione regionale. Alinea, Florence
Blečić I (2012) Costruzione degli scenari per la pianificazione. Franco Angeli, Milan
Blečić I, Cecchini A (2016) Verso una pianificazione antifragile: come pensare al futuro senza prevederlo. Franco Angeli, Milan
Bobbio L (2004) A più voci. Amministrazioni pubbliche, imprese, associazioni e cittadini nei processi decisionali inclusivi. Edizioni scientifiche italiane, Naples
Bonfiglioli S (ed) (1995) Il piano degli orari. Antologia di materiali per progettare ed attuare politiche pubbliche. Franco Angeli, Milan
Bonomi A (2002) La comunità maledetta. Viaggio nella conoscenza di luogo. Edizioni di Comunità, Turin
Borlini B, Memo F (2009) Ripensare l'accessibilità urbana. Cittalia ANCI Ricerche, Rome
Brunetta G, Caldarice O (2020) Spatial resilience in planning: meanings, challenges, and perspectives for urban transition. In: Filho L et al (eds) Sustainable cities and communities, pp 1–12. Springer International Publishing
Caniglia Rispoli C, Signorelli A (2001) L'esperienza del piano Ina-Casa: tra antropologia e urbanistica. In: Di Biagi P (ed) La grande ricostruzione. Il piano Ina-Casa e l'Italia degli anni cinquanta, pp 187–204. Donzelli Editore, Rome
Casciato M (2001) 'L'invenzione della realtà'. Realismo e Neorealismo dell'Italia degli anni cinquanta. In: Di Biagi P (ed) La grande ricostruzione. Il piano Ina-Casa e l'Italia degli anni cinquanta, pp 205–222. Donzelli Editore, Rome
Censis (2021) Rapporto annuale 2021. Franco Angeli, Milan
Choay F (2003) Espacements. Figure di spazi urbani nel tempo. Skira, Milan
Clementi A (2004) Mutamenti del contesto e ambivalenza urbanistica. Territorio, n. 28, pp175–179. Franco Angeli, Milan
Clifford S, Maggi M, Murtas D (2006) Genius Loci. Perché, quando e come realizzare una mappa di comunità. IRES—Istituto di Ricerche Economico Sociali del Piemonte, Turin
Comim F, Qizilbash M, Alkire S (eds) (2010) The capability approach. Cambridge Books, Cambridge University Press
Comune di Milano (2020) Milan 2020, Adaptation Strategy. Open document to the city's contribution. https://www.comune.milano.it/documents/20126/7117896/Milano+2020.+Adaptation+strategy.pdf/d11a0983-6ce5-5385-d173-efcc28b45413t?t=1589366192908
Consonni G (2019) Carta dell'Habitat, Confcooperative Habitat—La Vita Felice, Milan
Comune di Milano (2017) Documento di Visione Strategica. Scali ferroviari. Comune di Milano, Milan
Corboz A (1985) Le territoire comme palimpseste. Diogène, no. 121/1983, It. tran., Il territorio come palinsesto. Casabella no. 516, pp 22–27. Mondadori, Milan

de Certeau M (1980) L'invention du quotidien. Union générale d'éditions, Paris

Dupuy JP (2011) Per un catastrofismo illuminato. Quando l'impossibile è certo. Medusa Edizioni, Milan

Dupuy JP (1996) Prefazione. In: Pucci P (ed) I nodi infrastrutturali: luoghi e non luoghi metropolitani. Franco Angeli, Milan

Estreguil C, Dige G, Kleeschulte S, Carrao H, Raynal J, Teller A (2019) Strategic green infrastructure and ecosystem restoration, EUR 29449 EN, publications office of the European Union. Luxembourg. https://doi.org/10.2760/36800

Galuzzi P, Magnani M, Solero E, Vitillo P (2019) Residual urban spaces and news communities of social practices. TRIA 2:31–50

Galuzzi P, Solero E, Vitillo P (2020) Alpine space fragilities. A Research Line. Territorio 92:181–184

Gasparrini C (2012) Città da riconoscere e reti eco-paesaggistiche. Eco-Logics—PPC no. 25–26, Pescara

Gasparrini C, Terracciano A (2017) Dross city. Metabolismo urbano e progetto di riciclo dei drosscape. List, Trento

Gehl J (2011) Life between buildings: using public space, 1st ed. 1971, Island Press

Gregotti V (1985) Morfologia, materiale. Casabella, no. 515, pp 2–3, Mondadori, Milan

IPCC—Intergovernmental Panel on Climate Change (2022) Climate change 2022: impacts, adaptation, and vulnerability. In: Pörtner HO, Roberts DC, Tignor M, Poloczanska ES, Mintenbeck K, Alegría A, Craig M, Langsdorf S, Löschke S, Möller V, Okem A, Rama B (eds) Contribution of working group ii to the sixth assessment report of the intergovernmental panel on climate change. Cambridge University Press

Ippolito F (2012) Tattiche. Il Melangolo, Genoa

Jacobs J (1961) The death and life of great American cities. Random House, New York

Lavorato A, Galuzzi P, Vitillo P (2020) 8 Racconti di Milano. Ance, Milan

Lefebvre H (1976) La produzione dello spazio. Moizzi, Milan

Lindblom CE, Cohen DK (1979) Usable knowledge: social science and social problem solving. Yale University Press, London

Lynch K (1960) The image of the city. MIT Press, Cambridge

Manzini E (2021) Abitare la prossimità: Idee per la città dei 15 minuti, Egea

Mareggi M (2011) Ritmi urbani. Maggioli Editore, Sant'Arcangelo di Romagna (RN)

Mei P (2015) Il tempo della simultaneità nel progetto urbano. Tra permanenza e mutazione. Maggioli Editore, Sant'Arcangelo di Romagna (RN)

Montedoro L (2013) Una scelta per Milano. Scali ferroviari e trasformazione della città. Quodlibet, Macerata

Olmo C (2018) Non sempre le mitografie hanno ragione. In: Olmo C, Bonifazio P, Lazzarini L (eds) Le case Olivetti a Ivrea. L'Ufficio Consulenza Case Dipendenti ed Emilio A. Tarpino, Il Mulino, Bologna

Palermo PC (2022) Il futuro dell'urbanistica post-riformista. Gangemi, Rome, forthcoming

Panerai P, Castex J, Depaule JC (1981) Isolato urbano e città contemporanea. CittàStudi, Milan

Pasqui G (2008) Città, popolazioni, politiche. Jaka Book, Milan

Pasqui G (2017) Urbanistica oggi. Piccolo lessico critico. Donzelli, Rome

Pasqui G, Sini C (2020) Perché gli alberi non rispondono. Lo spazio urbano e i destini dell'abitare. Jaca Book, Milan

Pucci P, Vecchio G (2019) Enabling mobilities. Planning tools for people and their mobility. Springer, Berlin

Rupprecht Consult (2019) Guidelines for developing and implementing a sustainable urban mobility plan, Second Edition, Cologne

Russo M (2011) Città mosaico. Il progetto contemporaneo oltre la settorialità. Clean Edizioni, Naples

Secchi B (1989) Un progetto per l'urbanistica. Einaudi, Turin

Secchi B (2000) Prima lezione di urbanistica. Laterza, Rome-Bari

Secchi B (2001) I quartieri dell'Ina-Casa e la costruzione della città contemporanea. In: Di Biagi P (ed) La grande ricostruzione. Il piano Ina-Casa e l'Italia degli anni cinquanta, pp 149–160. Donzelli Editore, Rome

Sica P (1992) Storia dell'Urbanistica. L'Ottocento. Laterza, Rome-Bari

Steinitz C (ed) (2012) A framework for geodesign: changing geography by design. ESRI Press, Redlands, California

Stoker V, Chotray G (2009) Governance theory and practice: a cross-disciplinary approach. Palgrave Macmillan, London

Vitale T (2009) Introduzione: elogio del possibilismo. In: Vitale T (ed) Politiche possibili, pp 14–20. Carrocci Editore, Rome

Vitillo P (2021) Città contemporanea e prossimità. Urbanistica Informazioni, no. 295, pp 124

Wade W (2012) Scenario planning. A field guide to the future. John Wiley & Sons, Hoboken, New Jersey

Welter WM (2005) In-between space and society. In: Risselada M, van den Heuvel D (eds) Team 10 1953-81. In search of a utopia of the present. Netherlands Architecture Institute, Rotterdam

WHO—World Health Organization (2006) The world health report 2006: working together for health. World Health Organization, Geneva

Zucchi C (2018) Tra progetto e contesto. Città Sostenibili e Resilienti. Equilibri 1:27–31

Zucchi C (2020) Una città (non) è un albero. In lode di un disegno urbano just-out-of-time. Equilibri, no. 2, pp 421–430

Zucconi G (2001) La città dell'Ottocento. Laterza, Rome-Bari

Fig. 5.1 Diagram of the preliminary structure of the project for the Sforza-Policlinico station. The definition of the project for the reorganisation of the surface areas of the Sforza-Policlinico station began with the identification of three environments, known as the 'rooms' of the urban project. The first room, closest to the exit from the M4, develops the theme of the open-air museum; the second focuses on the requalification of the parking lot in front of the Basilica di San Nazaro in Brolo; the third room, closest to the exit from the M3, develops the theme of promoting modern architectural heritage

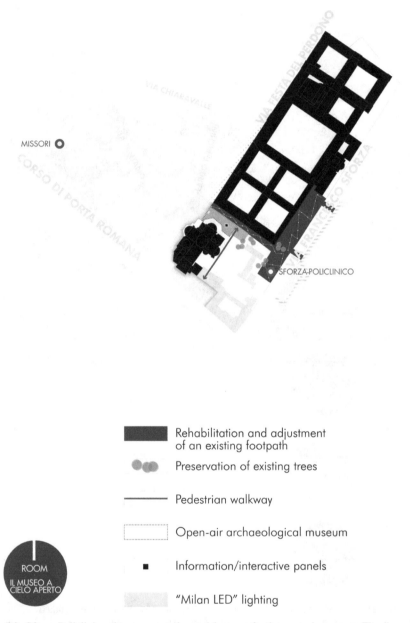

Fig. 5.2 Sforza-Policlinico, the campus station and the room for the open-air museum. The diagram identifies the priority actions for reorganising the public space between the State University and the hospital, interpreting their relative spaces and Via F. Sforza as spaces in which to promote existing archaeological remains, and the Renaissance and Baroque architecture fronting it and the Naviglio which is to be reopened in the future

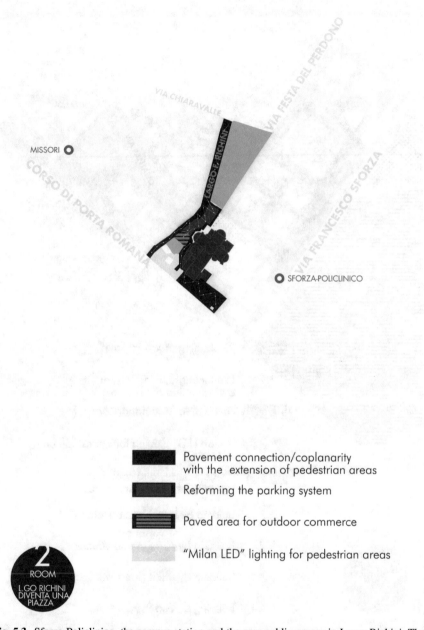

Fig. 5.3 Sforza-Policlinico, the campus station and the new public square in Largo Richini. The diagram identifies the priority actions for reorganising the public space that surrounds the asilica di San Nazaro in Brolo, proposing the redesign of draining paving and a reduction in areas accessible to motorised vehicles to promote pedestrian uses and connections

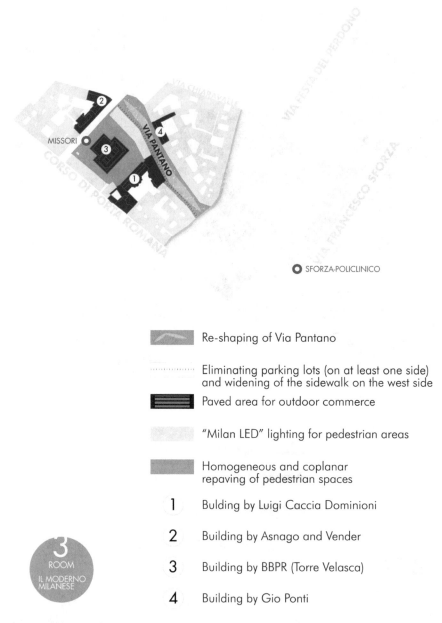

Fig. 5.4 Sforza-Policlinico, the campus station and the room of modern Milanese architecture. The diagram identifies the priority actions for reorganising the public space along Via Pantano to rediscover the work of well-known Milanese architects located near this axis, currently of scarce landscape value

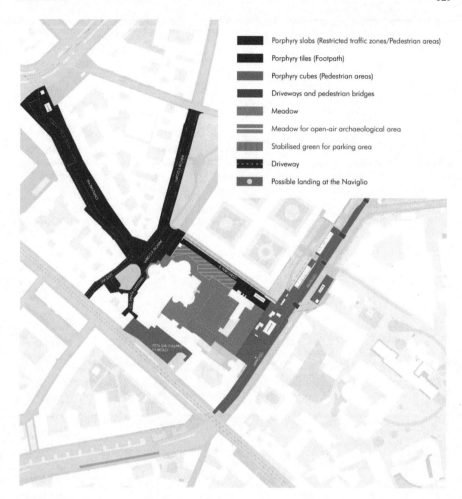

Porphyry slabs (Restricted traffic zones/Pedestrian areas)

Porphyry tiles (Footpath)

Porphyry cubes (Pedestrian areas)

Driveways and pedestrian bridges

Meadow

Meadow for open-air archaeological area

Stabilised green for parking area

Driveway

Possible landing at the Naviglio

Fig. 5.5 Design study of materials. The diagram summarises the design guidelines for repaving the areas around the Sforza-Policlinico station. The objective is to create a continuity between spaces, to facilitate the use of the urban environments created following the opening of the new station, and to facilitate the passage of users (not only the most fragile) along the connection between the Missori station on the M3 line and the new M4

Printed in the United States
by Baker & Taylor Publisher Services